Human Population Dynamics:
Cross-Disciplinary Perspectives

In human populations, biological, social, spatial, ecological and economic aspects of existence are inextricably linked, demanding a holistic approach to their study. Many undergraduate and postgraduate courses now emphasise the value of studying human populations using theoretical frameworks and methodologies from different traditional disciplines. *Human Population Dynamics* introduces such frameworks and methodologies whilst demonstrating how changes in human population structures can be addressed from several different academic perspectives. As such, the book contains contributions from world-renowned researchers in demography, social and biological anthropology, genetics, biology, ecology, history and human geography. In particular, the contributors emphasise the lability of many population structures and boundaries, as viewed from their area of expertise. This text is aimed at undergraduate students, graduates and academic researchers from any academic discipline which considers human populations.

HELEN MACBETH is an Honorary Research Fellow and retired Principal Lecturer in the Department of Anthropology at Oxford Brookes University. Her research interests have focused on the interaction of biological and cultural processes and their effects on human biology. She was Chair of the Biosocial Society and is currently Chair of the European section of the International Commission on the Anthropology of Food.

PAUL COLLINSON is an Honorary Research Fellow and former Lecturer in the Department of Anthropology at Oxford Brookes University. His research interests encompass the anthropology of development, European anthropology, social change, the anthropology of conflict and the anthropology of food.

THE BIOSOCIAL SOCIETY SYMPOSIUM SERIES

Series editor: Dr Catherine Panter-Brick, University of Durham

The aim of the Biosocial Society is to examine topics and issues of biological and social importance and to promote studies of biosocial matters. By examining various contemporary issues and phenomena, which clearly have dimensions in both the social and biological sciences, the Society hopes to foster the integration and inter-relationships of these dimensions.

Previously published volumes

1. Famine *edited by G.A. Harrison*
2. Biosocial Aspects of Social Class *edited by C.G.N. Mascie-Taylor*
3. Mating and Marriage *edited by V. Reynolds and J. Kellett*
4. Social & Biological Aspects of Ethnicity *edited by M. Chapman*
5. The Anthropology of Disease *edited by C.G.N. Mascie-Taylor*
6. Human Adaptation *edited by G.A. Harrison*
7. Health Interactions in Less-developed Countries *edited by S. J. Ulijaszek*
8. Health Outcomes: Biological, Social & Economic Perspectives *edited by H. Macbeth*
9. Biosocial Perspectives on Children *edited by C. Panter-Brick*
10. Sex, Gender and Health *edited by T.M. Pollard and S. B. Hyatt*
11. Infertility in the Modern World *edited by G.R. Bentley and C.G.N. Mascie-Taylor*
12. Hunter-Gatherers *edited by C. Panter-Brick, R.H. Layton and P. Rowley-Conwy*

Volumes 1–8 are available from Oxford University Press.

Human Population Dynamics: Cross-Disciplinary Perspectives

Edited by

HELEN MACBETH
and
PAUL COLLINSON

Oxford Brookes University

CAMBRIDGE
UNIVERSITY PRESS

CAMBRIDGE UNIVERSITY PRESS
Cambridge, New York, Melbourne, Madrid, Cape Town, Singapore,
São Paulo, Delhi, Dubai, Tokyo, Mexico City

Cambridge University Press
The Edinburgh Building, Cambridge CB2 8RU, UK

Published in the United States of America by Cambridge University Press, New York

www.cambridge.org
Information on this title: www.cambridge.org/9780521004688

First published 2002

A catalogue record for this publication is available from the British Library

Library of Congress Cataloguing in Publication Data

Human population dynamics: cross-disciplinary perspectives / edited by
Helen Macbeth and Paul Collinson.
   p. cm. – (Biosocial Society symposium series)
Includes bibliographical references and index.
ISBN 0 521 80865 0 ISBN 0 521 00468 3 (pbk.)
1. Population. 2. Sociobiology. 3. Social evolution. 4. Demography.
I. Macbeth, Helen M. II. Collinson, Paul, 1969– III. Series.
HB871 .H874 2002 304.6 – dc21 2001052627

ISBN 978-0-521-80865-1 Hardback
ISBN 978-0-521-00468-8 Paperback

Contents

Contributors

R.A. ATTENBOROUGH
Department of Archaeology and Anthropology, Hope Building,
A.N.U., ACT 0200, Australia.

M.J. BAMSHAD
Department of Pediatrics, University of Utah Health Sciences
Center, Salt Lake City, UT 84112, U.S.A.

J. BERTRANPETIT
Unitat de Biologia Evolutiva, Facultat de Ciencies de la
Salut i de la Vida, Universitat Pompeu Fabra, Doctor Aiguader 80,
08003 Barcelona, Spain.

F. CALAFELL
Unitat de Biologia Evolutiva, Facultat de Ciencies de la
Salut i de la Vida, Universitat Pompeu Fabra, Doctor Aiguader 80,
08003 Barcelona, Spain.

J.I. CLARKE
Department of Geography, University of Durham, Science
Laboratories, South Road, Durham DH1 3LE, U.K.

P. COLLINSON
Department of Anthropology, School of Social Sciences
and Law, Oxford Brookes University, Headington, Oxford
OX3 0BP, U.K.

M.E. DIXON
Department of Human Genetics, University of Utah Health
Sciences Center, Salt Lake City, UT 84112, U.S.A.

G.A. HARRISON
Institute of Biological Anthropology, University of Oxford,
58 Banbury Road, Oxford OX2 6QS, U.K.

P.R.A. HINDE
Department of Social Statistics, University of Southampton,
Southampton SO17 1BJ, U.K.

L.T. HUMPHREY
Department of Palaeontology, The Natural History Museum,
Cromwell Road, London SW17 5BD, U.K.

L.B. JORDE
Department of Human Genetics, University of Utah Health
Sciences Center, Salt Lake City, UT 84112, U.S.A.

P. KUNSTADTER
Institute for Health Policy Studies, University of California,
1388 Sutter Street, 11th Floor, San Francisco, CA 94109, U.S.A.

R.H. LAYTON
Department of Anthropology, University of Durham,
43 Old Elvet, Durham DH1 3HN, U.K.

H.M. MACBETH
Department of Anthropology, School of Social Sciences
and Law, Oxford Brookes University, Headington,
Oxford OX3 0BP, U.K.

J.M. NAIDU
Department of Anthropology, Andhra University, Visakhapatnam,
Andhra Pradesh, India.

B.V.R. PRASAD
Department of Anthropology, Andhra University, Visakhapatnam, Andhra Pradesh, India.

B.B. RAO
Department of Anthropology, Andhra University, Visakhapatnam, Andhra Pradesh, India.

C.E. RICKER
Department of Human Genetics, University of Utah Health Sciences Center, Salt Lake City, UT 84112, U.S.A.

E.K. ROUSHAM
Department of Human Sciences, Loughborough University, Loughborough, Leics LE11 3TU, U.K.

R. SMITH
Cambridge Group for the History of Population, Sir William Hardy Building, Department of Geography, Downing Place, Cambridge CB2 3EN, U.K.

W.S. WATKINS
Department of Human Genetics, University of Utah Health Sciences Center, Salt Lake City, UT 84112, U.S.A.

Foreword

Evolutionary biology became gradually transformed through the twentieth century from what was termed 'typological thinking' to 'population thinking'. This happened from the increasing recognition that differences between individuals were as important as similarities. Variation contains the raw material of evolution and variation can exist only in populations. The transformation harmonised the relationship of evolutionary biology with other population based biologies: genetics, ecology and epidemiology, for example. More significantly, however, in humans it also established potential connections with some of the social sciences and especially sociology, social anthropology, human geography and social psychology, all of which are fundamentally population based areas of knowledge.

Population can thus be seen as a bridge between the 'two cultures' of natural sciences and the humanities. This was most effectively recognised by J.W.S. Pringle at Oxford, who was instrumental in establishing there the field of Human Sciences, essentially based on the analysis of human population structures from all perspectives. In the past twenty-five years the field has developed dramatically.

The concept of population is not, however, one without difficulties. Even the definition of a population can be a major problem. Human groups rarely exist as discrete, more or less uniform entities even at any one time, and they have never existed over time. Typically they intergrade, often very gradually. Then while one most frequently thinks of populations in some spatial context, they also exist in vertical dimensions such as social class and caste, and in ecological and economic terms reflecting natural environmental heterogeneity. Clearly a population defined by one criterion can be made up of a number of others defined by another, and boundaries are rarely the same. Thus a spatial population can be made up of a number of genetic or social populations each with a variety of resource bases, some of which are shared with other spatial populations. Obviously it is always necessary to define a population according to the particular purposes of study or report. But there is a universal: all natural populations share in things, be they space, genes, resources or culture. The concept of sharing is thus a fundamental feature of population definition, and variation in the degree of sharing allows a continuous scale of population hierarchies to be recognised.

From such considerations it is evident that demography lies at the core of the Human Sciences. The size, growth and distribution of any population unit depends on the fundamentals of comparative birth and death, and the issue of 'sharing' typically revolves around some form of movement whether it be spatial migration or social mobility or movement through the generations by inheritance. Every component of the Human Sciences has demographic components and most of the interconnections between these components pass through demography. In the Human Sciences, however, one sees demography in a rather wider context than is usual and incorporating other methodologies than strictly demographic ones so as to achieve a full understanding of population dynamics.

The situation is evaluated in detail in this exceptional book. It achieves its coherence by focusing on demographic approaches, but both in detail and at a general level it indicates the relevance of those approaches to examining a diversity of biosocial phenomena. It will be invaluable to all Human Scientists.

G. AINSWORTH HARRISON

Preface

It is clear that there are many perspectives on human populations and that these are studied within different disciplines, whose terminology and literature are becoming increasingly specialised. Furthermore, the features and conditions of the populations are always changing and the processes of this are viewed and reported upon from different angles. What is required, therefore, for a complete understanding of the complex dynamics of human populations is an appreciation of the work of specialists in different disciplines, something which this volume provides. It is anticipated that the book will be useful to all who consider human populations in their research or their studies, and particularly those who are undertaking degree courses in human sciences, anthropology, demography, human geography, ecology and human biology. The book will also have a wider appeal for the more general reader who wishes to advance their understanding of human population dynamics by adopting a wide-ranging disciplinary focus.

This volume arises from the fifteenth workshop of the Biosocial Society, which was held at the Pauling Human Sciences Centre, Oxford, in 2000. The Society thanks Ros Odling-Smee and the Centre for accommodating the conference. The help with arrangements for and at the workshop by the team of postgraduates from Oxford Brookes University is gratefully acknowledged, and the Society is thanked for providing finances for the workshop. The Editors also wish to thank Chris McDonaugh for reading the book and making helpful suggestions, Jennifer Jay for work on the glossary, index and referencing, and finally Geoff Harrison for writing the Foreword.

H.M.M. and P.S.C.

The publisher has used its best endeavours to ensure that the URLs for external websites referred to in this book are correct and active at the time of going to press. However, the publisher has no responsibility for the websites and can make no guarantee that a site will remain live or that the content is or will remain appropriate.

1

Introduction: the framework of studying human population dynamics

HELEN MACBETH AND PAUL COLLINSON

Humans can be studied from a variety of academic perspectives. In some biological disciplines the focus may be on parts of individuals, perhaps even molecular parts, while in others humans are considered zoologically as an entire species, *Homo sapiens*, in the order *Primatii*. Some biologists are concerned with the physiologically functioning body of an individual, while others are interested in world-wide human diversity, distribution and population groupings. Among the social sciences, the focus is also frequently on the group, but here the emphasis tends to be upon the social and cultural factors which underpin the way such phenomena as societies, communities and ethnic groups are constructed, delimited and defined. There are also the behavioural sciences, which utilise both biological and social information in relation to the study of the group, the individual or elements in the behaviour of the individual. Recognition of the number of perspectives on these population variables emphasises the need to study material across the boundaries of traditional academic disciplines and several of these perspectives are introduced in this volume.

The twentieth century began with observers of the human condition integrating, in what we now recognise to be a confused way, their ideas on the biological and social nature of humans. During the century the fragmentation of all disciplines grew, something exemplified most particularly in the divergence between biological and social pursuits of knowledge. This in turn fuelled many bitter debates between some biologists and some social scientists who were insufficiently informed on the theories and use of language of their antagonists. During the last few decades, however, the recognition of the interaction of social and biological processes has resulted in an increasing number of university courses which perhaps claim or aim at integrating disciplines; these are certainly multidisciplinary if not successfully interdisciplinary. At the beginning of a new century, we feel that 'integration' of older disciplines may not be an appropriate ambition, but that the need is for those involved in studying any aspect of human groups to be well acquainted with at least the fundamental ideas of other disciplinary approaches. While this volume is generally concerned with the demographic variables of fertility, mortality and migration, its principal aims are to provide an introduction to the different approaches to the study of how human populations change over time, and to emphasise how components of each perspective should be considered by those interested in any aspect of the processes of human population dynamics.

Disciplines which include the study of human groups are diverse, and because of the interaction of the subject matter, there can be intricate debates on nomenclature. However, the editors of this volume feel that a simplified introduction to some of them is required here. Demography is, in essence, the calculation of population numbers and population changes, and brings to bear many perspectives on the causes, patterns and consequences of fertility, mortality and migration. Since births, deaths and migration are central variables in the work of practitioners of the other disciplines which together form the 'Human Sciences', a full understanding of demography involves consideration of these other disciplinary approaches. One can start with anthropology, as the word means the study of humans. Within anthropology there are many sub-disciplines and these overlap in so many ways that one can only distinguish them in very general terms. The palaeoanthropologists study prehistoric humans, but hominid evolution is also a significant part of physical anthropology and is considered within the study of human population genetics. In practice, there is no difference today between physical and biological anthropology, although the reason that many departments changed their name to the latter was to emphasise a greater understanding of molecular diversity and to reduce the emphasis on measuring morphological characteristics, such as human skulls and physique. The labels ecological anthropology and medical anthropology might appear to be self-defining, and yet there lurks big disciplinary diversity within each. Cultural anthropology and social anthropology are both titles for disciplines concerned with human cultural and social behaviour, structures, organisations and institutions. While both include a strong focus on the study of the interaction between different human groups, to enter into discussions of their differences is beyond the scope of this introduction. In general, however, under the title of cultural anthropology there is a wider span of sub-disciplines, and use of that title is more common in America than in Europe. The science of diverse cultural practices and materials can also be called ethnology, and ethnography is the descriptive writing about a population and its practices. The subject matter of sociology is similar to that of social and cultural anthropology, although the emphasis here is more upon the study of social problems and their origins. Historically, sociological research has also tended to be focused more on the populations of industrialised countries than on those of the less economically developed regions of the world, which were studied more commonly by cultural and social anthropologists. This boundary between the disciplines has become far less distinct in recent years. Whereas human geographers were traditionally concerned with spatial distribution, their discipline now includes a strong emphasis on the study of the interaction between human beings and the environment, in terms of both the impacts of human activity upon the environment and the way in which the environment influences human organisation and behaviour. Zoologists, especially primatologists, include humans in their studies, and population geneticists estimate distributions of gene frequencies within and between human groups. Although the main focus of ecologists may frequently be on the non-human environment, little of this is unaffected by human activity and vice versa. Many other disciplines consider human populations, for example history, economics and linguistics, but traditionally what has been insufficiently emphasised within each discipline is the need to cross academic boundaries into the terminology and material of other disciplines. This, we argue, is essential in relation to the understanding of human population dynamics.

Taxonomists attempt to divide all species into constituent 'populations', and the human species is no different in this regard. However, this endeavour is far more difficult when applied to humans than to other species. From a biologist's point of view, when considering isolated and divergent breeding groups of the same or very similar non-human species (on the Galapagos Islands, for example), one even can describe the geographic boundaries of that breeding population. Where the boundaries cannot be identified, for example with more migratory or mobile species, the clear delimitation of a 'breeding population' may be impossible. If there is no barrier to mating, some miscegenation should be expected, blurring any boundaries based on biological characteristics. As has been suggested from the observation of several closely related species, a barrier to successful mating can arise through behavioural differences. When applying these concepts to humans, the complexity of the problem is greatly compounded.

Although human beings world-wide belong to one species and theoretically reproduction is biologically possible between any fertile male and any fertile female, there clearly are barriers that prevent all possible matings actually taking place. Firstly, there are geographic barriers, one of which is simply distance: people who do not meet cannot mate. However, humans are an increasingly migratory species and all such barriers are crossed by some individuals sooner or later. The shorter and longer movements of individuals and groups across geographic space have given rise, over evolutionary time, to clinal distributions of gene frequencies, making the division of humans into genetically discrete populations impossible. So, in studying human populations from a biological perspective, it is clear that we cannot be limited by the biologists' concept of the breeding, or 'Mendelian', population (for a definition see the Glossary).

As well as physical barriers, humans have layers of social ideas about appropriate marriage and mating partners. In terms of the ways human groups are constituted, these represent more important barriers than those rooted in simple geography, since they are strongly related to other concepts of group identity and vary widely between different cultures and societies. As well as the social and political significance of such social constraints on marriage patterns, they also affect future patterns of gene frequencies.

Demographers and other enumerators refer to a population as all those within any boundaries, which are usually divisions between administrative regions and need not have any effect whatsoever on mating patterns, past or present. In contrast, practitioners from some social science disciplines are not so interested in establishing any 'objective' definition of a population. Far more important for them is to establish the ways in which human groups define themselves and those around them, usually involving reference to the inter-action of complex social, cultural and/or ethnic characteristics. Social anthropologists and sociologists have identified the ways in which groups generate boundaries between one another, boundaries which, although highly porous, represent the basis of the collective identities of those who are, or choose to be, located within them. The facets of social identity applied to this 'boundary formation process' are highly varied, and may include, among many others, a common lan-guage, specific sets of symbols and rituals, laws of prescription and proscription, and other rules of behaviour, even modes of dress or cuisine. Commonly, a combination of many such aspects of social life is used. Such ideas are relatively easy to understand when ap-plied to national identities: nation states deliberately exploit collec-tive symbols of nationhood, such as national anthems, flags, parades, etc., in order to generate a sense of belonging among their citizens. This process, some might argue, is now being applied at a supra-national level by the European Union. However, it is also present in the emotional adherence to associations and groupings of much smaller scale.

Whether demographically, geographically, genetically, socially or politically, one can perceive, even if not precisely delineate, levels of populations within populations. Harrison and Boyce (1972) describe the patterns within populations as 'structures' with biological and cultural characteristics. Such structures can be as all-inclusive as the whole human species or as small-scale as the nuclear family. Between these two extremes the word 'structure' seems particularly appropri-ate, as one can perceive a hierarchical pattern from the perspective of any discipline: geographically, it might be shown from home to town or village to country to continent, etc.; socially, from household to community to 'ethnic' or language group, etc.; administratively, from parish to district or province to nation state to federation of states, etc. Because of the existence of these structures, as smaller associa-tions cluster within larger groupings and these in turn can be seen to be more or less united under some other umbrella, one must be clear about the level or scale that one is considering in any discussion (Clarke 1972). To give an example, a move of residence from Paris to Rome might be called a migration between two nation states, while viewed from a different perspective it is residential mobility within the European Union. Similarly, at one level it is possible to estimate the gene frequencies of those dwelling in the Indian subcontinent, while at another level the gene frequencies of different castes may be compared. In the discussion of culture, demography or gene frequencies, one must identify the specific group or be aware that one may be generalising about a larger association of communities.

However, social boundaries exist at every level and many individuals operate in many different spheres at once. Moreover, social boundaries usually exhibit a 'segmentary' characteristic, at times drawing groups together, at other times being a means of separation, thereby creating complex 'patterns' forged by geographic, historic and cultural factors. In this way, establishing a set of 'objective' analytical criteria by which one can understand the various ways human populations are constructed becomes highly problematic. 'Communities', 'social groups', 'ethnic groups', 'societies' or even 'nations' utilise essentially very similar processes of self-construction and definition, and often alter their nature and form greatly over time or in different contexts. Today's ethnic group may well become tomorrow's nation state, and vice versa. It is clear that separating one group from another, or even defining the point at which one group ends and another begins, is an extremely complex process, and one which renders the very concept of 'population' – at least when used in its biological sense – virtually meaningless. The most important point for our purposes is that social boundaries, by their very fluidity, are by no means immutable. The most important point to stress in this cross-disciplinary volume is that all boundaries can be crossed and in this way neither demographic processes nor genetic inheritance become encapsulated within any population, however delimited. Whatever disciplinary lens is applied to the problem – be it demography, human geography, human biology, cultural anthropology or even town planning – any student of human population dynamics must always take into account what people themselves perceive their own 'population' to be. (For an ethnographic illustration of the complexity of these processes see Kunstadter, Chapter 9.)

It seems clear that from the angle of any of the disciplines mentioned above, partly flexible structures can be perceived in the ways in which humans associate. However, while the continuity of some features may well be highly significant in the self-definition of some associations, in their academic description it is not uncommon that insufficient attention has been given to the changes that take place in all societies, cultures, environments and gene frequencies. The patterns of the structures themselves are labile and interdependent. Furthermore, the complexity and mutability of human groupings can be perceived from each of the disciplinary perspectives and thus the actuality is an even more intricate web of interdependent and changing dimensions that defy any of the traditional attempts at taxonomy. In our view an introduction to these different perspectives is a basic step in the study of human populations, a recognition which stimulated this book.

The structure of this volume

The contributors were chosen because of their expertise in different disciplines and each chapter provides a different academic approach to the study of humanity. Furthermore, as the components of all population structures tend to change over time, the mechanisms of change are also discussed. While the benefits of multidisciplinary study have already been defended in this introduction, the challenge for all cross-disciplinary volumes is to achieve a level of coherence through the chapters. In this volume, the vehicles of population dynamics, fertility, mortality and migration provide that thread of continuity, as they are developed within the different perspectives.

The first of the disciplinary approaches to be introduced is that of demography, the study of population numbers and how these change. Hinde's presentation of demographic analysis (Chapter 2) starts from first principles, as the processes and specialist terms are explained with exceptional clarity. He includes a definition of population which, for demographers, can be based on any observable characteristic. After showing how fertility, mortality and migration are the mechanisms for numerical change this chapter focuses on the first two, fertility and mortality, and the theoretical approach to how they are analysed and recorded in a 'closed population' (p. 20). Debates about population growth are in the public arena these days, but frequently with little precision. This introduction to the analytical methods of demographers is, therefore, prerequisite to the study of human population dynamics. Many of the terms and concepts explained by Hinde reappear in subsequent chapters.

In Chapter 3, Clarke demonstrates the overlap between demography and geography in regard to an interest in population numbers and increase. His chapter starts with some statistics about world population growth over the second half of the twentieth century, but he explains the relevance of disaggregating the figures, showing how most of the world's population growth has occurred in the less developed countries (LDCs). He identifies diversity between LDCs, which he links to economic conditions. As a geographer, however, he introduces the relevance of spatial distributions and the changing situation between the continents. So, whereas Hinde had kept his theoretical discussions to closed populations, Clarke is very much concerned with the effects of migration. Of particular relevance to his argument is rural depopulation and the growth of cities and mega-cities, many of which have grown up within easy reach of the seas to benefit economically from maritime communication and rich coastal plains. The concentration of so much of the world's population in coastal regions makes them vulnerable to climate change. His chapter ends with projections about urban growth and its impact on the human condition.

The next perspective is social, provided by the social anthropologist, Layton (Chapter 4), who starts with clear definitions of population, community and society. These definitions are essential for the development of his discussion about peasant communities. He includes information on past debates on the processes of social construction and social change. After interesting cross-disciplinary reference to the interaction between sociocultural developments and genes, his chapter concentrates on the relevance of different social rules, especially those of inheritance, to family size and population dynamics, particularly in peasant communities. The interaction of social rules and other aspects of population dynamics are clearly exemplified in this chapter.

The description of the biologist's interpretation of the word 'population', given by Layton, seems at first to need no further explanation in by Bertranpetit and Calafell (Chapter 5) on the geneticists' perspective on population dynamics. Yet, as their discussion develops, it is clear that the understanding of human biological diversity has progressed to a far more complex level with the information accruing from the Human Genome Project. The processes by which the frequencies of genes change within a population have long been recognised as mutation, natural selection, genetic drift and migration. As natural selection and genetic drift are caused by differences in fertility and mortality, the relevance of these demographic variables is again exemplified. By linking the understanding of the processes of molecular and population change to the new wealth in human DNA information, which is available and still accumulating world-wide, a new comprehension of the evolutionary dynamics of populations, of their admixtures and subdivisions, of their founding origins and the rate of their growth, is now accessible. The chapter demonstrates the way in which molecular evidence can be used to throw light on the ancestral history of human groups, the results of such work further complicating the concept of population.

The next three chapters also introduce three different disciplinary perspectives, but the first has more emphasis on fertility, the second on mortality and the third on migration and social mobility. Smith, a historical demographer, begins Chapter 6 with a review of the *Essay on the Principle of Population* by Thomas Malthus (1803), who compared his contemporary European society with 'more uncivilised parts of the world' in regard to what he termed 'positive checks' on fertility. This chapter shows how academic analyses of fertility developed since Malthus, and the significance of nuptiality patterns and household formation. The relevance of social factors to the patterns is clear. While comparison is first drawn between western European societies and those east of a line from Trieste to St Petersburg, the chapter continues to discuss significant historical differences between examples in India and those in China. Reference is also made to China's modern one-child policy. Further comparison is made with African societies.

The concept of homeostasis in population size is raised here and is central to the discussions in the final chapter by Attenborough.

Rousham and Humphrey (Chapter 7) concentrate on child mortality and survival, using information from epidemiology and palaeopathology. The chapter starts with comparisons of the high child mortality in LDCs because of infectious disease, especially when combined with malnutrition, and the much lower child mortality in developed countries. Child mortality not only depends upon the interactions of genetic and non-genetic factors, but also affects demographic, social and genetic structures of populations, and the relevance of medical conditions, cultural practices and other world inequalities is shown. Among relevant cultural practices, gender discrimination is shown in the excess mortality of female children and infants, and it is argued that economic development does not necessarily eliminate practices of sex selection. The chapter goes on to report on what is known about the causes of child mortality in European populations before industrialisation. Improvements in England and Wales during the twentieth century are referred to as well as strategies to reduce child mortality in the LDCs.

In relation to how the study of DNA material can throw light on past migrations and subsequent social mobility, especially of females, the next chapter gives details of a research project in which various types of DNA were analysed. Jorde *et al.* (Chapter 8) provide a general introduction to the three successive waves of immigration into the Indian subcontinent which, they argue, have left identifiable markers in the genetic makeup of different groups within the region. While it is recognised that today the caste system is much more complex than outlined here, the authors introduce the basics of this social stratification, pointing out that there are also other elements in the social structure that may be described as 'tribal'. The proposition by Bertranpetit and Calafell (Chapter 5) that detailed DNA information can be used to throw light on past mobility is demonstrated in the evidence presented by Jorde *et al.* Again greater female mobility, this time through the caste system, is shown. The social practices, including hypergamy and concubinage, in the society of south India, with its complex subdivisions, have had an effect on the genetic structure of the population, and the DNA evidence is used to throw light on the social practices. Their conclusions are consistent with archaeological data, and historical and contemporary information. Hypotheses are discussed and this chapter provides clear evidence that the information from studies of molecular DNA when linked to other archaeological, historical, demographic and social data can be used to elucidate past human population dynamics, but when viewed without the other perspectives is hard to interpret.

The final two chapters demonstrate the ability of the practitioner of 'Human Sciences' to cross many disciplinary boundaries in their discussion of human population dynamics. Kunstadter (Chapter 9) draws on his own research, which spans four decades, on the Lua' and Hmong highland minority groups in Thailand. Here the variables of mortality, fertility and migration are all shown to have changed in pattern over the second half of the twentieth century. This chapter provides a rich ethnographic insight into the two populations, showing how external changes, government actions, malaria control, road building, markets, etc., have had very different effects on the population structures of these two tribal or ethnic groups. Underlying social, cultural and religious systems, Kunstadter explains, have played a highly significant role in the way the population structures and demography of the Lua' and the Hmong have developed quite differently. Not only are fertility and mortality patterns discussed with data and ethnographic detail, but also the difference between the populations in migration patterns and post-migration community cohesion. Agricultural and economic conditions, both locally and, in the case of Hmong, internationally, are considered. Thai official programmes today tend not to discriminate between the different hill tribes, and while the majority population views these tribes as backward, it uses them with their colourful traditional clothing to attract tourism and to entertain official visitors. A new picturesque identity is imposed upon them. The importance of cross-disciplinary and longitudinal research is demonstrated as the discussion includes contemporary dynamics in the concepts of ethnic boundaries and how each of these populations has reacted quite differently to events. With cultural boundaries blurring in different ways, with Lua' marriage patterns including non-Lua' and Hmong marriage patterns remaining endogamous but on an international scale, the biological, social and spatial definitions of what is a 'population' discussed in earlier chapters are all shown to be insufficient. However, the population dynamics caused by concepts of identity are undeniable.

The editors have decided to end the book with the broad-ranging chapter by Attenborough. Here the ecological perspective is introduced in the context, appropriately, of many other disciplinary perspectives. Attenborough's debate focuses on whether population homeostasis, or very slow growth, occurred in some ecological settings. He reviews information gained from the study of non-industrial populations in remote and difficult environmental circumstances, referring to some populations that have survived with only slow, if any, population growth, and a few cases where their demise can be studied from archaeological remains. His search is for clear evidence of homeostasis over the long term, and this proves to be elusive, although strongly indicated. One feature that adds to the difficulties of finding a 'closed system' in which to study homeostasis, is that even remote populations are found to be far less isolated than first imagined. So once again the labile, or at least porous, nature of population boundaries, despite natural barriers, is an issue.

The contemporary and future dynamics of human populations

Mentioned briefly in more than one place in the following chapters are some features of contemporary human conditions that are having far-reaching effects on the demographic, social, economic and biological structures and the spatial distributions of humans, and which will have significant future repercussions as well. Readers will no doubt identify many other issues and perspectives on the human condition that are either under-represented or totally neglected in this volume. For example, very little is mentioned about the effects of trade and other ever-changing economic processes on the demographic variables of fertility, mortality and migration, although these effects have become important topics in many human sciences courses. There are only a few references to the effects of HIV/AIDS or environmental change. We do not feel that this volume could encompass full discussions of all these issues; However, it would be lax of the editors if we did not refer to at least some of them here.

Whilst we live on what has been termed a 'human dominated planet', changes in environmental and climatic processes during the late twentieth century, some of which can be attributed directly to this domination, have brought into sharp relief the inherent vulnerability of our species. The effects of global warming are now being felt in many regions of the world, and will have undoubted implications for human population dynamics in future decades. As Clarke points out (Chapter 3), it is likely that the trend towards increasing concentration of populations will accelerate as marginal and low-lying coastal areas become less habitable, and the already enormous refugee problems which beset many regions of the world will be exacerbated by further mass migrations.

The geographically disproportionate impact of environmental change is only likely to increase the suffering caused by malnutrition among the poorest people in the world, and to further highlight the increasing divisions between 'developed' and 'developing' regions. Whilst the effects of malnutrition are considered by Rousham and Humphrey (Chapter 7), the political issues of food, water and agri-cultural policies, both governmental and non-governmental, are also of great contemporary significance to human population dynamics. The disproportionate terms of trade (and 'aid') between countries and regions, the increasing power of transnational corporations and the export of free market economic policies are all creating enormous changes for the lives of millions of people, throughout our increas-ingly interconnected planet. However, as the post-colonial era has given way to the age of 'globalisation' and the 'information revolu-tion', especially in industrialised countries, the most serious problems facing the human species are being faced predominantly by those for whom this shift in paradigms is largely meaningless. The impact of AIDS, for example – now the leading infectious disease in the world – has been felt predominantly in the developing world, where its implications for the dynamics of populations will be felt for many generations to come.

In terms of infection rates alone, the impact of HIV has been startling in its severity. This is particularly true in sub-Saharan Africa, where the enormous social, economic and political problems have been amplified by the age ranges of those infected. This has led to a very rapid shift in demographic structures such that, for example, in many areas the burden of childcare has shifted from parents to grandparents and/or older siblings. As revealed in the Biosocial Society's 2001 Symposium, *Learning from HIV/AIDS: transdisciplinary perspectives*, the pandemic has highlighted the need for the cross-disciplinary approach to human population dynamics. It is particularly exemplified in attempts to reduce the incidence of this disease in the absence, so far, of an effective vaccine or other medical intervention. Behavioural change has seemed to be the only route to improvement, and public intervention programmes in many countries, even in the poorest regions of the world, have demonstrated some success. An estimated 23 million deaths due to AIDS world-wide by 2001 makes its demographic, biological and socioeconomic effects at least comparable with historical outbreaks of the bubonic plague or the influenza pandemic of 1918, and the urgency of bringing to bear as many disciplinary perspectives as possible on the condition all the more pressing.

In contrast to the demographic, genetic, social and economic effects of infectious disease in general and AIDS in particular in poorer regions, in industrial nations it is the increasing rate of geriatric survival that is affecting population structures. The increasing proportion of the population who are no longer economically active, due to age and retirement, is a departure from historical situations. This too has many social, spatial, economic, medical and ethical implications for national governments and other agencies, and the effects have been amplified by the marked decline in fertility rates in most advanced economies in the world. As couples increasingly choose to place their careers and material wealth above marriage and childbirth, many 'developed' nations are now exhibiting negative population growth rates for the first time in their histories (see Attenborough, Chapter 10). The end of the welfare economy is already a feature of some post-industrial societies, as governments realise that the costs associated with these two demographic trends are beginning to outstrip the resources available.

We could discuss many more issues of contemporary concern for the study of human population dynamics, but what we hope to have demonstrated is that we are living in an era of unprecedented change, the understanding of which will require increasing cross-disciplinary insight and collaboration. For this it is necessary for practitioners to be able to translate theoretical concerns, analysed from different perspectives, into practical ways of ameliorating some of the problems that present generations are donating to their inheritors.

Conclusion

This volume has brought together some of the perspectives out-lined above. As humans have mammalian biology, reproduction and mortality, as well as highly complex social, cultural and economic institutions, in differing spatial and ecological settings, with different climates, vegetation and pathogens, all of which are regularly chang-ing, those who study the human condition in the twenty-first century should become more informed on a wide range of disciplinary ap-proaches to these variables than the late-twentieth century tendency towards academic specialism generally allowed.

The considerable similarity in some objectives and problems of the social and the biological scientists who study modern human groups should be highlighted rather than avoided. The definition of the grouping is one such problem. In this volume the definition of the human population is shown to be complex, as not only do the social, demographic, genetic and spatial structures within and between pop-ulations change, but the boundaries of what is considered a population are themselves labile. This being the case, then the holistic study of those groups of people and how their characteristics, numbers, condi-tions and structures change requires co-operation between scientists from different disciplines. That is the rationale for this book.

Acknowledgements

The authors thank Alan Collinson for suggestions which have been incorporated into this chapter.

References

Clarke, J.I. (1972). Geographical influences upon the size, distribution and growth of human populations. In *The Structure of Human Populations,* ed. G.A. Harrison and A.J. Boyce, pp. 17–31. Oxford: Oxford University Press.

Harrison, G.A. and Boyce, A.J. (ed.) (1972). *The Structure of Human Populations.* Oxford: Oxford University Press.

Malthus, T.R (1803). *An Essay on the Principle of Population.* [Volume 2 of *The Works of Thomas Robert Malthus*, ed. E.A. Wrigley and D. Souden (1986). London: William Pickering.]

2

Demographic perspectives on human population dynamics

ANDREW HINDE

Introduction

This chapter describes a formal demographic perspective on human population dynamics. It first attempts to summarise the way in which human population dynamics are treated in the more technical and theoretical demographic literature. The next section considers some demographic fundamentals, including population structure (especially the age and sex composition) and the three components of population change: fertility, mortality and migration. The third part looks at some of the formal models which demographers have developed to help understand population change. These models make several assumptions in order to simplify a complex reality. One of these is that migration is zero: populations with zero migration are said to be *closed*. An attraction of this is that, if migration can be ignored, simple relationships exist between fertility, mortality, the population growth rate and the age structure.

In the fourth section population dynamics in the short to medium term are considered. The age and sex structure of a population is itself a dynamic feature, containing a record of the population's past fertility, mortality and migration. Moreover, the future age–sex structure is determined by past and current events. Discussion of these aspects of population dynamics leads naturally in the fifth section to a consideration of population momentum, or what is sometimes, inaccurately, called the 'demographic time bomb'. The origins of population momentum are explained.

Finally, in the sixth and seventh sections long-run population dynamics are explored in the context of the demographic transitions in

Europe and contemporary Africa. The relationships between fertility, mortality and population growth enable closed populations to be placed in 'fertility–mortality space'. Placing historical and contemporary populations in this space reinforces the idea that long-run population growth is naturally very slow. It also allows comparisons to be made of the demographic transition in different cultures.

Some demographic fundamentals

Demography is about analysing the growth (or decline) of human populations and changes in their structure. We cannot analyse population growth without some straightforward and unambiguous way of enumerating the members of a population, and this can only exist if we can work out who is, and who is not, to be included in the enumeration.

To demographers, a *population* consists of any group of persons who can be delimited on the basis of some observable characteristic. The most common way is to define the population on the basis of residence within a given geographical area. Thus we speak, for example, of the 'population of England' as being all those persons normally resident in England. However, residence is not the only criterion. We could delimit populations on the basis of tribal affiliation: the Luo population of Kenya, for example, using both geographical residence and membership of a particular tribe as criteria. Necessary features of characteristics used to delimit a population are that they be observable and well defined, so that we can use them to say whether any individual person is or is not a member of the population at a given time. Notice that although these characteristics are used to distinguish separate populations, they themselves are inclusive: they are shared features which bind the members of a distinct population together.

In order to be able to analyse population growth and change, it is essential to be able to identify the processes by which persons enter or leave populations. It is fortunate for demographers that populations only change in size because of a limited, countable, range of events. Consider, for example, the population of a particular country at some time, say 1 January 2000, which we might call P_{2000}. Then the population of that country on 1 January 2001, P_{2001}, is equal to P_{2000} plus the number of births during the year 2000, minus the number of deaths, plus the number of people who migrate into the country during the year, minus the number of persons who migrate out. The difference between the number of births and the number of deaths is known as *natural increase* (or *decrease* if deaths exceed births) and the difference between the number of immigrants and the number of emigrants is known as *net migration*.

Analysing changes in the size of human populations, therefore, involves the analysis of the processes by which births, deaths and migration events come about. The process which produces births is known as *fertility*; the corresponding process which results in deaths is called *mortality*. The three processes of fertility, mortality and migration are known as the components of population change.

Demographers are interested particularly in the intensity with which these events occur in a particular population. Since a large population will tend to generate more events than a small population, the absolute numbers of, say, births and deaths in a particular time period are of limited use as measures of this intensity. Therefore, demographers use what are called *rates*. A demographic rate is a ratio of events of a particular type, for example deaths, to the number of persons exposed to the risk of experiencing that type of event. Thus the *crude death rate* is equal to the number of deaths in a given period divided by the 'average' population during that period. It is normally important that the events and the population 'exposed to risk' correspond. That is, we need to make sure that the persons exposed to risk really are at risk of experiencing the event in question, and that we do not include events in the numerator which do not occur to persons in the denominator. There are some exceptions to this, one being the *crude birth rate* (number of births in a given period divided by the 'average' population during that period) in which men, who do not give birth, are included in the denominator.

Because people are not all identical, all human populations have a *structure*. By *population structure* demographers mean the distribution of various characteristics across the members of a population. The characteristics most commonly considered by demographers are sex and age. Certainly, these will be the most important variables so far as this

chapter is concerned. Other variables by which a population structure may be defined include genetic make-up, socioeconomic classification, occupation, educational attainment, place of birth, ethnic
affiliation, etc.

A very important feature of populations is that the intensity of the
components of population change varies according to people's characteristics. Thus, different subgroups within a population will have
different risks of experiencing births, deaths and migration. People
aged over 60 years have a higher risk of dying than do teenagers,
for example. For this reason, demographers tend to work with rates
specific to particular subgroups. The most commonly used of these
are age-specific rates, but in principle, rates can be calculated specific to any subgroup of interest, for example, as defined by variables
discussed above in regard to structure.

Formal demographic models of population change

If migration is ignored (that is, if we assume a closed population),
simple relationships exist between a population's fertility, its mortality and its rate of growth, and between these three variables and its
age structure. These relationships can be described mathematically,
and this permits the construction of elegant demographic models
of human population dynamics. Two sets of relationships are of
great interest in a closed population (Figure 2.1). The first is that
linking fertility and mortality to the rate of growth. The second is
that linking the rate of growth and mortality to the population's age
structure.

Let us consider the first of these. A convenient way to understand
how the relationship works is to imagine the population of a remote
island. Suppose that on this island there are 100 men and 100 women,
and suppose that each man is 'married' to one woman so that we have
100 couples. By 'married' we simply mean that the couple is in a more
or less stable sexual relationship, whether this is legally formalised or
not. One plausible measure of the rate of population growth is the
total number of children that these 100 couples produce. Clearly,
if this is greater than 200, then the next generation will be larger

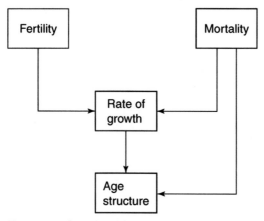

Figure 2.1. Relationships in a closed population.

than the present generation, and the population will grow. If it is less than 200, the population will decline. The total number of children produced by the 100 couples can be divided by 100 to give the average completed family size. If the average completed family size exceeds 2, then the population will grow. Demographers often refer to the average completed family size as the *total fertility rate* (TFR). If, then, on our remote island the TFR is 4, each woman will produce four children. Suppose that half of these are boys and half girls. The result will be that our 100 couples will have 400 children, 200 sons and 200 daughters. If the same TFR is maintained for subsequent generations, then 800 grandchildren will be born, 1,600 great-grandchildren, and so on. The population will therefore double in size every generation. This, of course, is what Thomas Malthus meant when, in his *Essay on the Principle of Population* (1798), he described populations as increasing in a 'geometrical ratio'.

The TFR is, in fact, the sum of a set of *age-specific fertility rates* (ASFRs). Recall the definition of a demographic rate in the previous section. The ASFR is defined as the number of births to women of a given age in a given year divided by the number of women of that age, i.e. events divided by those exposed to risk. It measures the number of births that the 'average' woman can expect to have in a year while she is at that age. So, for example, the ASFR for a woman aged 20 years last birthday is a measure of the number of children the average woman has in the year between her 20th and 21st birthdays. In practice, this will be less than 1 in almost all human populations, as most women in most populations will have no children at all between their 20th and 21st birthdays. If we sum the ASFRs at all ages from the youngest age at which women bear children to the oldest age, the result will be an estimate of the number of children which the average woman will produce during the whole of her childbearing period, or, in other words, the TFR.

Population growth, therefore, depends on fertility. However, it matters whether the children are boys or girls. To see this, imagine the populations on three remote islands, A, B and C. Each of these islands contains a population of 100 couples. Suppose that on all these islands each couple has exactly four children, i.e. the TFR is equal to 4. On island A equal numbers of boys and girls are born. However, on island B three boys are born for every girl, and on island C three girls are born for every boy. Island A exhibits the doubling every generation that we have already described. On island B the original 100 couples produce 300 sons and 100 daughters, whereas the 100 couples on island C produce 100 sons and 300 daughters. Thus in the first generation, the population of both islands doubles. However, in the next generation things become more interesting. Assuming that the women continue to have four children each, then on island B only 400 children will be born in the third generation, whereas on island C, each of the 300 daughters of the first generation will produce four children, making a total of 1,200 grandchildren (300 grandsons and 900 granddaughters). Of course, achieving this would involve polygyny or monogamy with extra-marital childbearing. For the population of island B to match the growth of island C, each of the daughters of the first generation would have to produce 12 children. Even supposing that polyandry were widespread, as there are, after all, three males for every female in the second generation, it is most unlikely that this will be achieved. Indeed, the highest reliably recorded fertility in a human population is around ten children per woman among the Canadian Hutterite population during the 1920s and 1930s. The key point to take away from this stylised example is that the rate of population growth depends on the number of girls born rather than on the number of boys.

In practice the sex ratio of births among human populations varies rather little; it averages 105 or 106 boys per 100 girls. This implies that a TFR of 4 will lead to a population not quite doubling itself every generation. It is likely that an awareness that, so far as population growth is concerned, it is daughters that matter, almost certainly lay behind the female infanticide practised by certain tribal populations in the past.

This mention of sex-selective infanticide (see Rousham and Humphrey, Chapter 7) leads neatly into the final factor determining the rate of population growth in a closed population. This, of course, is mortality. And what matters here is the chance that a daughter will survive long enough to have children herself. Returning to the remote islands, let us suppose that on islands D and E the sex ratio of births is 105 boys per 100 girls. Suppose, however, that on island D four in every five girls born survive to reproductive age, but that on island E only one in every two girls does this. On island D, the 100 women in the original generation will bear 400 children, of which 195 will be daughters $(400 \times 100/205)$. Only four-fifths of these $(195 \times 0.8 = 156)$ survive to reproductive age. These 156 daughters will bear 624 (156×4) children in total, of which 304 will be girls $(624 \times 100/205)$. Of these, 80% will survive to reproductive age, resulting in a population of 243 granddaughters. The female infant and child mortality has reduced the rate of population growth substantially. This is even more strikingly illustrated by island E, in which only half of those born survive to reproductive age. It can be shown that the population of island E will decline over the generations and eventually die out.

For population growth, the mortality of females matters more than that of males, though male mortality is not completely irrelevant. Moreover, even for females, it is only mortality up to and during the childbearing years that is of interest. Thus, to analyse population growth, we need to measure age-specific female mortality up to about age 50 years. *Age-specific death rates* (ASDRs) can be calculated in a way similar to that of ASFRs, by dividing the deaths in a given year to persons of a specific age by the population of that age. A set of ASDRs provides a complete description of the mortality experience of a population. In particular, ASDRs can be used to draw up a table of the probability that a person will survive to at least a given age. This table is known as a life table, and an example is shown in Table 2.1, using

Table 2.1. *Conditional probabilities of death and probability of survival to different ages, females in England and Wales, 1901–10*

Age x	Conditional probability of death before next specified age	Probability of survival to at least age x
0	0.1174	1.0000
1	0.0689	0.8826
5	0.0173	0.8218
10	0.0106	0.8076
15	0.0143	0.7990
20	0.0173	0.7876
25	0.0208	0.7740
30	0.0265	0.7579
35	0.0334	0.7378
40	0.0414	0.7131
45	0.0529	0.6836
50	0.0705	0.6475
55	0.1001	0.6018
60	0.1374	0.5416
65	0.1942	0.4672
70	0.2983	0.3764
75	0.4116	0.2641
80	0.5436	0.1554
85	1.0000	0.0709

The conditional probability of dying before the next specified age is the chance that a woman will die prior to the next specified age given that she is still alive at age x. So, for example, the chance that a woman aged exactly 40 years will die before her 45th birthday is 0.0414 (just over 4%). Clearly, this conditional probability for the oldest specified age x must be 1.0000, as everyone eventually dies.

Source: Woods and Hinde (1987: 33).

data which relate to the female population of England and Wales in the decade 1901–10. Under this mortality regime, for example, a woman had about an 80% chance (probability 0.7990) of surviving to her 15th birthday, and about a 65% chance (probability 0.6475) of surviving to her 50th birthday. Associated with the probabilities of survival to given ages are conditional probabilities of death within a particular age range. These conditional probabilities are expressions of the chance that someone will die within an age range given that they are alive at the start of that age range. An important special case is the infant mortality rate, which is the probability that a baby will die before his or her first birthday. From Table 2.1, it can be seen that the infant mortality rate for girls in England and Wales in 1901–10 was 0.1174. In other words, almost 12% of girls born during this period did not survive until their first birthday. Although the mortality of males is not so important for understanding population growth, it is worth pointing out that in most human populations, the mortality of males exceeds that of females at all ages. Infant and child mortality is discussed further by Rousham and Humphrey (Chapter 7).

The stylised example described in this section shows that population growth depends on three things: the average number of children born to women, the sex ratio of births and the chance of a female child surviving to reproductive age. One measure of population growth which incorporates all of these is the *net reproduction rate* (NRR). The NRR is approximately equal to the TFR multiplied by the proportion of births that are girls multiplied by the probability of a woman surviving to the mean age at childbearing. The mean age at childbearing varies a little between populations, but is between 27.5 and 30 years in most cases. Therefore, suppose that a population had a TFR of 4.0, that 105 boys were born for every 100 girls, and that the mortality of females was described by Table 2.1. The chance that a woman will still be alive at exact age 25 years is 0.7740, and at age 30 years it is 0.7579. So the probability of survival to the mean age at childbearing (27.5–30 years) may be estimated at, say, 0.76. The NRR is, then, calculated as $4 \times (100/205) \times 0.76 = 1.48$. The NRR measures the size of the next generation relative to the size of the present generation. So an NRR of 1.48 means that the next generation will be 48% larger than the current one.

The second important relationship is that between the rate of growth, the mortality and the age structure of a population. The number of persons alive at any age x, at a particular time, in a closed population, is the product of two factors: the number born x years ago and the probability of survival to age x. If the crude birth rate and the ASDRs in the population are constant over time, it is possible to show that the population will grow in size at a constant rate (see e.g. Coale 1987). If this rate is r per year, it is then possible to show that the number of births x years ago is equal to the number of births in the current year multiplied by a factor equal to e^{-rx}, where $e = 2.718$. Multiplying these births by the probability of survival to age x, taken from a life table representing the population's mortality, will produce the number of persons now alive aged x (for more details, see Hinde 1998). Notice that for the age structure, mortality at all ages matters, not just at ages below 50 years.

The assumptions of a constant birth rate and a constant set of ASDRs are, of course, rather restrictive, and are unlikely to be met in any real populations. However, the benefits of having simple relationships between fertility, mortality, the rate of growth and the age structure are considerable, and so idealised populations of this sort are often considered by demographers. Populations embodying all the assumptions made so far in this chapter are called *stable*. The special case where a stable population has a growth rate of zero (i.e. the absolute number of births and the absolute number of deaths are the same) is called a *stationary population*. In a stationary population, the number of persons alive at each age is constant from year to year.

Examples of the age structures of stable populations with different rates of growth are shown in Figure 2.2, drawn as population pyramids. Characteristically, growing populations have pyramids with wide bases, reflecting their youthful age structure (Figure 2.2A). Declining populations have pyramids which are narrower at the base than in the 'middle' (Figure 2.2C), reflecting the fact that in more recent years there have been progressively fewer children born. Stationary populations have so-called 'rectangular' pyramids (Figure 2.2B).

Population dynamics in the short to medium term

The foregoing discussion has shown how a population's age structure is determined by its mortality rate and its rate of growth. Since the rate

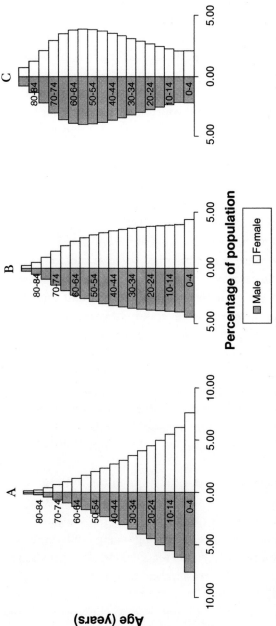

Figure 2.2. Population pyramids in a growing, a stationary and a declining population. All three populations have the mortality rates given in Table 2.1. Population A is growing at an annual rate of 2%, population B is stationary, and population C is declining at an annual rate of 2%.

of growth is, in turn, determined by fertility and mortality through the net reproduction rate, then we can see that a population's fertility and mortality will alone determine its age structure. This is true even when we relax the assumption that fertility and mortality are constant over time. It can be proved mathematically (see e.g. Lopez 1967) that a population's age structure is determined only by its past fertility and mortality. In other words, if one subjects two closed populations to the same sequence of fertility and mortality rates for long enough, they will come to have the same age structure, no matter how different their initial age structures were. The initial age structures will, in a sense, be 'forgotten'. The only exceptions are certain rather weird age structures, such as a population with no-one aged under 50 years, which is, of course, doomed to die out.

Therefore a population's age structure is a dynamic thing. It carries within itself an historical record of that population's past fertility and mortality experience. This may be seen most conveniently by considering a real example. Figure 2.3 shows the population pyramid of England and Wales in 1995. It can be seen that there was a relatively large number of persons aged 45–49 years. These people were born between 1945 and 1950, during which there was a short, but intense, baby boom associated with demobilisation after World War II. A longer baby boom occurred during the 1960s, reflected in the large number of people aged 25–34 in 1995. During the early 1970s, however, fertility declined rapidly (the TFR fell from over 2.5 in the mid-1960s to about 1.7 in the mid-1970s). The small number of people aged 15–19 years (born between 1975 and 1980) is the legacy of the low fertility of this period.

The current age structure of a population will also influence its future development in the short to medium term. To say this is not to contradict the statement that ultimately populations 'forget' their previous age structures: this 'forgetting' occurs only in the long term (100 years or more). Over shorter periods, a population's present age structure will exert considerable influence over its future development. This is, perhaps, best seen by considering the phenomenon of population momentum, sometimes described as the 'demographic time bomb'.

Age (years)

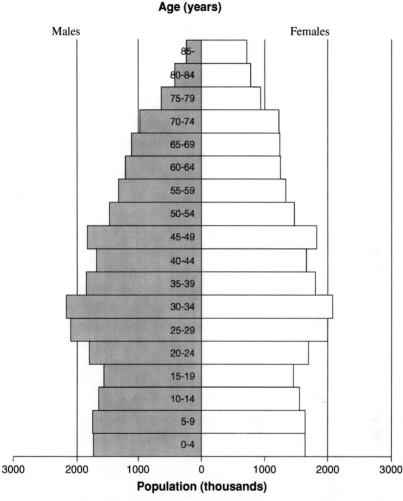

Figure 2.3. Population pyramid of England and Wales in 1995.
Source: Office for National Statistics (1997: 1).

Population momentum

In order to understand population momentum a few preliminaries are
needed. The first involves the distinction between what demographers
call *period* phenomena and the experience of different groups of real
people. Demographers call groups of real people *cohorts*, and normally

refer to them by their year(s) of birth. For understanding population growth, it is the cohort perspective which is most relevant, because what matters is how many children (particularly daughters) a real generation of women will produce. An NRR of 1.0 denotes that the generation is just replacing itself, and is known as *replacement level*. Notice that this is not the same as women having, on average, two children. For if the average completed family size is 2.0, then the fact that slightly more boys are born than girls and mortality at ages up to 50 years will mean that the NRR is less than 1.0. Indeed, with the mortality of England and Wales in 1901–10 (Table 2.1), in which 76% of girls born survive to the mean age at childbearing, a TFR of 2.0 will produce an NRR of only 0.74 ($2.0 \times (100/205) \times 0.76$). This implies a population decline of 26% in each generation. For a population with this level of mortality to have an NRR of 1.0 would require the average woman to bear 2.7 children. Clearly, therefore, 'replacement level fertility' is greater than two children per woman. In low mortality populations, however, in which over 95% of girls born survive to childbearing age, it is not much greater, being about 2.1.

Now, consider a population in which fertility falls abruptly to replacement level. This fall takes place in a given period, but it affects different cohorts of women at different ages. The cohorts which are of childbearing age in this period will have had many of their children in previous years. For example, those women who are aged 35–39 years when fertility falls to replacement level will have been bearing children for 20–25 years at the previous higher fertility rates, and will only have 10–15 years left to bear children at replacement level fertility. Their average completed family size will, therefore, largely reflect the previous higher fertility, and will be greater than replacement level. Only when fertility has been at replacement level for long enough for a cohort to spend all its childbearing years experiencing ASFRs corresponding to replacement level, will that cohort produce a generation which is the same size as the cohort itself, that is, having an NRR of 1.0 (see Table 2.2 for an example).

However, even when a population has an NRR of 1.0, there will still be population momentum remaining, for the current age structure influences the future development of the population. At the point when the NRR reaches 1.0, the people born during earlier periods,

Table 2.2. *Illustration of population momentum*

| | Age-specific fertility rates by period | | | | | | | | | | | | |
Age group	1970–74	1975–79	1980–84	1985–89	1990–94	1995–99	2000–04	2005–09	2010–24	2015–19	2020–24	2025–29	2030–34
15–19	0.0675	0.0675	0.0675	0.0675	0.0675	0.0675	0.045	0.045	0.045	0.045	0.045	0.045	0.045
20–24	0.2025	0.2025	0.2025	0.2025	0.2025	0.2025	0.135	0.135	0.135	0.135	0.135	0.135	0.135
25–29	0.204	0.204	0.204	0.204	0.204	0.204	0.136	0.136	0.136	0.136	0.136	0.136	0.136
30–34	0.1035	0.1035	0.1035	0.1035	0.1035	0.1035	0.069	0.069	0.069	0.069	0.069	0.069	0.069
35–39	0.039	0.039	0.039	0.039	0.039	0.039	0.026	0.026	0.026	0.026	0.026	0.026	0.026
40–44	0.012	0.012	0.012	0.012	0.012	0.012	0.008	0.008	0.008	0.008	0.008	0.008	0.008
45–49	0.0015	0.0015	0.0015	0.0015	0.0015	0.0015	0.001	0.001	0.001	0.001	0.001	0.001	0.001
Period TFR	3.15	3.15	3.15	3.15	3.15	3.15	2.1	2.1	2.1	2.1	2.1	2.1	2.1
TFR for cohort aged 45–49 years in period	3.15	3.15	3.15	3.15	3.15	3.15	3.1475	3.1275	3.0625	2.89	2.55	2.2125	2.1
Birth years of cohort aged 45–49 years in period							1953–57	1958–62	1963–67	1968–72	1973–77	1978–82	1983–87

The period TFRs are synthetic measurements which denote the average completed family size that would be obtained by a group of women who, at each age, had the age-specific fertility rates for that period. In this example, fertility falls abruptly on 1 January 2000 from a **TFR** of 3.15 to a **TFR** of 2.1 (replacement level). Estimates of the completed fertility of birth cohorts are obtained from the age-specific fertility rates by summing diagonally. Thus, for example, the 1953–57 birth cohort was aged 15–19 years in 1970–74, 20–24 years in 1975–79, and so on, and so its completed fertility may be estimated by summing the relevant diagonal elements and multiplying by five (because we are using five-year age groups). The resulting cohort TFRs for successive birth cohorts are shown. Not until the 1983–87 birth cohort, which completes its childbearing in 2030–34, is replacement-level fertility reached.

when the NRR was greater than 1.0, will still be alive. Before the abrupt fall in fertility, the population was growing. Therefore, its age structure will be youthful, resembling that of Figure 2.2A, with a large proportion of young people. In the years after the decline of fertility, these young people will grow older, producing a relatively large proportion of the population being of childbearing age. This will inflate the number of children born relative to the total population, thus raising the birth rate and delaying the arrival of 'zero population growth'. Eventually, of course, the age structure of the population will stabilise, but this is likely to take a few decades.

Population momentum applies to all changes in fertility and mortality. Abrupt 'period' changes in the components of population change tend to have delayed effects on population growth and on aspects of population structure. Simply reducing fertility to replacement level will not eliminate population growth immediately or even very quickly. The realisation of this was behind the Chinese government's one-child family policy. Because of the effects of population momentum, the Chinese government, which wanted to eliminate population growth as quickly as possible, saw the need to reduce fertility to below replacement level. Similarly, a population which has fertility below replacement level cannot expect to be able to increase population growth immediately by raising fertility to that level.

However, the metaphor of a 'time bomb' is inappropriate. The future development of a population's size and structure is predictable, given knowledge of its present age structure and future fertility and mortality rates. The reaction of a population to 'shocks' in the components of change may be delayed, but it is also gradual: the effect creeps up over several decades. It is this very delay which indirectly gives rise to the 'time bomb' effect, because the gradualness of demographic processes leads governments and other policy-making bodies to ignore them until they reach crisis proportions.

Population dynamics in the long term

In the short and medium term, human population dynamics for the demographer involve consideration, in the main, of the relationships between mortality, the rate of growth and the age structure of a population. However, in the 'long run' (i.e. many generations, often spanning several centuries), the interest in population structure becomes eclipsed by the importance of population size, and what matters most is the relationship between fertility, mortality and the rate of population growth.

From this perspective, it is also reasonable to ignore migration in practice, as well as in theory. Although there are many examples of populations which have grown or declined by large amounts through migration, over many generations natural increase almost invariably dominates net migration.

So, if we ignore migration, the population growth rate depends just on fertility and mortality. A convenient way of showing the growth rates which arise from different combinations of fertility and mortality is to draw a picture of *fertility–mortality space* (Figure 2.4). This takes the form of a chart with some measure of fertility on the vertical axis and a measure of mortality on the horizontal axis. Every population may then be placed somewhere in the resulting two-dimensional area. The scales of the axes can be chosen so that the normal range of human fertility and mortality experience is covered. The resulting space is divided into two by a line representing zero population growth. Above and to the right of this line populations will grow in size; below and to the left they will decline. The further a population is located from the zero growth line, the higher the rate of growth or decline.

In order to place real populations in this fertility–mortality space, measures of fertility and mortality need to be chosen. The crude birth and death rates could be used (see Woods and Hinde 1987). For consistency with the earlier discussion, however, we use here the TFR and the life expectancy at birth, the approach used by Livi-Bacci (1992: 21–3, 151). Figure 2.5 shows the result. The annual growth rates implied by different combinations of fertility and mortality can be worked out and lines drawn which join combinations of fertility and mortality leading to equal rates of growth (Livi-Bacci (1992: 21) calls these *isogrowth lines*). With a life expectancy at birth of 25 years, for example, a TFR of about five births per woman is required in order to achieve zero population growth. As mortality falls, and life

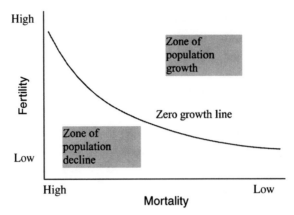

Figure 2.4. Fertility–mortality space.

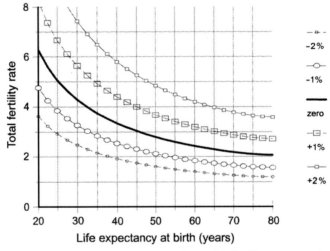

Figure 2.5. Population growth surface with fertility measured using the total fertility rate and mortality measured using the life expectancy at birth. The highest rates of growth are in the top right hand (north-east) corner, the most rapid population declines are in the bottom left hand (south-west) corner. Lines showing combinations of fertility and mortality leading to annual growth rates of 1 and 2% annual declines of 1 and 2% and zero growth are shown.
Source: Based on the design of Livi-Bacci (1992: 21).

expectancy rises, the TFR required to replace the population falls towards 2, though it never quite reaches that value. Figure 2.6 shows the current populations of a range of countries in contemporary Europe and Africa placed in this calibrated fertility–mortality space. Many contemporary African populations lie a long way above the zero growth line. This is a remarkable and, almost certainly, temporary phenomenon. For although short-term deviations are likely, sustained deviations above or below the line lead either to explosive growth which is checked by resource constraints (Malthus's famous positive check) or to extinction. For example, a population of 10 millions, if subjected to a growth rate of 2% per year, would grow to 74 millions in 100 years and 545 millions in 200 years. The same 10 millions, if they maintained a rate of decline of 2% per year for 100 years would fall to 1.35 millions, and after 200 years would number only 183,000 persons. Therefore, in the very long run, populations stay fairly close to the zero growth line.

Prehistoric populations, in so far as anything is known about their demography, appear to have had life expectancies at birth of 25 years or less. In order to have survived, therefore, they must have had TFRs of at least 5, occupying a zone at the top left-hand corner of the fertility–mortality space. The populations of contemporary developed countries cluster in a relatively small zone close to the zero growth line in the south-east corner of the chart (Figure 2.6). In the long run, therefore, population dynamics are about how populations move from the 'north-west' to the 'south-east' corner. This movement accompanies social and economic development, and is often called the *demographic transition*.

The theory of demographic transition

In the classical theory of the demographic transition, mortality normally declines before fertility, leading to a phase of rapid population growth. This is, in general, why many contemporary African countries are positioned far above the zero growth line: they are part way through this phase. Yet it is clear that this phase of

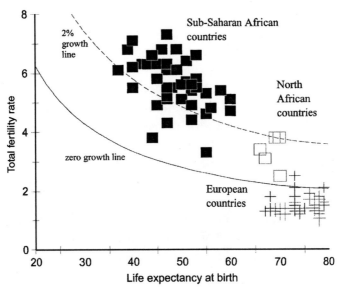

Figure 2.6. Location of certain contemporary populations in fertility–mortality space, using the total fertility rate to measure fertility and the life expectancy at birth to measure mortality. Source: United Nations: www.undp.org/popin/wdtrends/p98

rapid growth cannot be very long in historic terms, and that for the whole of the prehistoric era and most of the period from which documentary evidence survives, human populations have remained close to the zero growth line. However, what has also become clearer in recent years is that pre-industrial populations are not all clustered right in the 'north-west' corner of the fertility–mortality space. Instead they occupy a considerable area of the space. Between prehistoric times and the sixteenth and seventeenth centuries the populations of western Europe moved from the top left-hand corner of Figure 2.7 to a position towards the centre, with TFRs of between 4 and 5 and a life expectancy at birth of about 35 years. They did this largely by 'sliding down' the zero growth line, though the movement was probably uneven, and there were periods during which it was reversed, for example in the fourteenth century during the Black Death, and in subsequent epidemics. For further information on these matters see Livi-Bacci (2000).

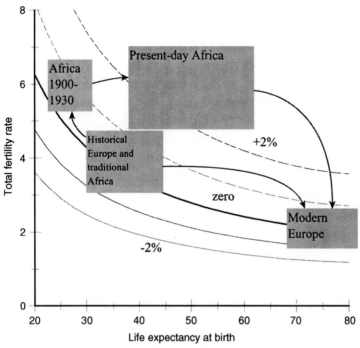

Figure 2.7. Illustration of the demographic transition in western Europe and sub-Saharan Africa.
Source: Adapted from Woods (1986: 12).

Since the seventeenth century, however, movement through fertility–mortality space has been much faster. It is this more rapid movement that is described as the demographic transition. Because the decline in mortality generally, though not invariably, precedes the decline in fertility, the movement is first predominantly in a horizontal direction (to the right) and later in a downward direction. However, because the distance to be travelled was relatively short, the movement never took the European populations very far from the zero growth line. For example, the annual rate of population growth in England was never more than 1.5% and England's rate of growth during the demographic transition was faster than that of most other European countries (Livi-Bacci 2000).

Why, then, have many contemporary African countries been experiencing growth rates of 2% or even 3% per year? The answer lies partly in their pre-transitional movements. It is likely that for several centuries prior to the twentieth, the position of many sub-Saharan African populations in fertility–mortality space was similar to that of western Europe in the sixteenth and seventeenth centuries, although the mechanisms by which fertility was regulated, such as lengthy breastfeeding and periods of abstinence from sexual inter-course following a birth, were different. However, during the late nineteenth and early twentieth centuries many African populations moved up and to the left. The reasons for this are related to their colonial history (see Mbacke 1994). The effect of this movement was that when the demographic transition began in earnest after World War II, these countries had much further to travel through fertility–mortality space to reach the position of modern industrial societies than had the historical populations of Europe. Because mortality in their case definitely declined before fertility, and declined rapidly as a result of the importation of modern medicine, rapid horizontal, leftward, movement has taken place before any substantial down-ward movement due to fertility decline. The result of this has been that many sub-Saharan African countries have found themselves pos-itioned at the beginning of the twenty-first century a long way above the zero growth line. However, as referred to in the last chapter, HIV is changing that position and the full effect of AIDS mortality is yet to be seen.

Conclusion

The aim of this chapter has been to describe some demographic perspectives on human population dynamics. The presentation has been rather theoretical, in order to highlight important relationships between fertility, mortality, a population's age structure and its rate of growth. Migration has been largely ignored, but is an important variable considered in later chapters.

It would be a shame if readers should get the impression from this chapter either that demography is a largely theoretical discipline, or that demography generally neglects migration, for neither of these statements is correct. However, it is, I think, fair to say that the discipline of demography has at its heart a 'formal' core, involving the measurement of rates and the elaboration of mathematical relationships surrounding population dynamics. It is work in this 'formal demography' which produces the perspective on human population dynamics which is peculiarly demographic, and which I have tried to describe in this chapter. Moreover, it is also the case that in this 'formal demography', migration is often ignored or assumed to be negligible.

However, surrounding the core of formal demography is a large corpus of more substantive work which demographers undertake to try to understand why the components of population change behave as they do. This work involves the consideration of the effects of a host of potentially explanatory factors, social, economic and cultural, on the components of change – and here migration is definitely included. In this chapter, that body of work has largely been pushed to one side, yet it forms by far the majority of the research that demographers actually do. Because the work of demographers overlaps with and draws from many other disciplines, its themes are echoed in other chapters in this volume, and the need for introductions to these other perspectives is demonstrated.

References

Coale, A.J. (1987). Stable populations. In *The New Palgrave: A Dictionary of Economics*, Vol. 4, ed. J. Eatwell, M. Milgate and P.K. Newman, pp. 466–469. London: Macmillan.

Hinde, P.R.A. (1998). *Demographic Methods*. London: Edward Arnold.

Livi-Bacci, M. (1992). *A Concise History of World Population*. Oxford: Blackwell.

Livi-Bacci, M. (2000). *The Population of Europe*. Oxford: Blackwell.

Lopez, A. (1967). Asymptotic properties of a human age distribution under a continuous net maternity function. *Demography*, **4**, 680–687.

Malthus, T.R. (1798). *An Essay on the Principle of Population* [Volume 1 of *The Works of Thomas Robert Malthus*, ed. E.A. Wrigley and D. Souden (1986)]. London: William Pickering.

Mbacke, C. (1994). Family planning programs and fertility transition in sub-Saharan Africa. *Population and Development Review*, **20**, 188–193.

Office for National Statistics (1997). *Mortality Statistics: Cause*, series DH2. No. 22. London: The Stationery Office.

Woods, R. (1996). Spatial and temporal patterns. In *Population Structures and Models: Developments in Spatial Demography,* ed. R. Woods, and P. Ress, pp. 7–20. London: Allen and Unwin.

Woods, R.I. and Hinde, P.R.A. (1987). Mortality in Victorian England: models and patterns. *Journal of Interdisciplinary History*, **18**, 27–54.

3

The growing concentration of world population from 1950 to 2050

JOHN I. CLARKE

Introduction

The beginning of the twenty-first century is a most opportune time to look back fifty years to see what has happened to the spatial patterns of world population distribution, and to try to look forward over a similar period to see what is likely to happen, especially as we are in the midst of an almost inexorable process of massive concentration of world population. That concentration must be examined within the context of other enormous demographic changes that have taken place since 1950 and are likely to take place before 2050. Naturally, the reliability of the retrospective analysis is greater than that of the prospective analysis, although I am relieved to note that most of my speculations and predictions made in comparable papers about three decades ago (Clarke 1971, 1972) have not been far from the mark. As Hinde (Chapter 2) has mentioned, the perspectives of demographers and geographers overlap, but the latter pay great attention to spatial distribution and migration.

Rapid world population growth

The mid-twentieth century was a demographic threshold when world population growth accelerated. At that time, the world population was just over 2.5 billion; according to the United Nations Population Division it reached 6 billion in October 1999. It had taken roughly a century and a half for world population to multiply two and a half times from 1 billion to 2.5 billion, but in the second half of the

Table 3.1. *World population growth*

Year	Total (bn)	Years to add 1 billion
1804	1	—
1927	2	123
1960	3	33
1974	4	14
1987	5	13
1999	6	12
2013 est.	7	14
2028 est.	8	15
2054 est.	9	26

Source of data: UN (1998a).

twentieth century the world population grew so rapidly that 3.5 billion were added to the world total, one billion being added during the last 12 years (Table 3.1).

Although towards the end of the twentieth century the time taken to add another billion to the world population was still diminishing, recent declines in fertility have meant that the rate and volume of population growth slowed down during the last years of the century to about 1.3% per annum or an annual increment of 78 million, from annual peaks of over 2% during 1965–70 and 86 million during 1985–90. However, with a broad base to the population pyramid and more than a billion persons aged between 15 and 24, considerable momentum in future world population growth is assured; thus, despite a projected drop in the annual increment during the first half of this century to 64 million during 2015–20 and then more sharply to 30 million by 2045–50, the United Nations 1998 medium-variant projection (UN 1998a) indicates that a total of 9 billion will be reached in 2054 (see Figure 3.1). Obviously, great weight should not be given to such a long-term projection, but over a period of 50 years the United Nations Population Division has a commendable record of world population projections, creating considerable confidence. Nevertheless, we should remind ourselves that we live in a period of ever-increasing social, scientific, economic and political change, including marked shifts and discontinuities (e.g. the Pill, AIDS, the

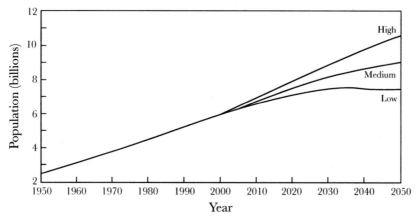

Figure 3.1. World population growth, 1950–2050. United
Nations Population Division 1998 estimates and high, medium
and low variant projections. Source: UN 1998a.

rise of electronic communications, the demise of communist governments, the attacks of 11 September 2001), which make all long-term
forecasting extremely difficult and speculative (e.g. UN 1998b).

One feature of future population growth about which there can be
little doubt is an increase in the number of older people. By 1998,
there were about 578 million people in the world aged 60 and over,
some 9 million being added to the total each year, an annual increment which is expected to rise to 14.5 million by 2010–15. By 2050 it
is projected that the percentage of older people aged 60 or more will
have risen from 10% to 22% of the world population. In the group
of so-called More Developed Countries (MDCs), the number of older
persons exceeded children under 16 for the first time in 1998, and
by 2050 it is expected that there will be more than twice as many
older people as children. We should recall that these ageing populations will have changing locational requirements in households and
residence.

Populations of states

The world population is merely an aggregate of the populations of
countries, which are all separately enumerated and whose number

has grown greatly since World War II. In 1945 there were only 72 independent states; by 2000 there were as many as 191 along with about 70 dependent states, all varying enormously in population size, characteristics, dynamics and policies to the extent that they are separate population systems. And there are growing pressures from national political movements for more to acquire independence, as for example in south-eastern Europe.

The scale of demographic variations between states is so great that the ten largest state populations, all with over 100 million people, account for 60% of the total world population, while at the other end of the scale dozens of states each have fewer than a million inhabitants. The population structures, distributions and dynamics of the many micro-states bear little comparison with those of the few macro-states, as the causes of demographic characteristics tend to vary with the size of population and of the territory occupied (Clarke 1972). We should note, for example, that the relationship between population and areal size is not direct; that areal size influences the hierarchical patterns of towns; that external migration rates of small areal units are higher than those of large units; that the diversity of demographic indices reduces with population size; and that the rates and structures of small populations are more variable in time and space. Thus the population of a small country like Singapore is not really comparable at all with that of China, still less with that of Asia or the world as a whole.

Since the middle of the twentieth century, there has been a vast improvement in the availability, accuracy and reliability of the population data of countries, especially the many Less Developed Countries (LDCs). This has taken place through a great increase in census-taking and surveys often supervised by the United Nations, so despite the persistence of major imperfections of data, especially in vital registration, there is firmer ground for making population projections.

For demographers, populations are usually spatially defined by the territories that they occupy rather than by their ethnic, national, cultural, genetic or demographic distinctiveness, and therefore they are identified by their political or administrative areas. Demography emerged to some extent as political arithmetic providing data and analyses for governments, but whether this administrative character will continue to prevail during the twenty-first century may be

Table 3.2. *Population growth of MDCs and LDCs, 1950–2050*

	Population (millions)			Av. pop. growth (% p.a.)	
	1950	1998	2050	1950–55	1995–2000
MDCs	812	1182	1155	1.3	0.3
LDCs	1709	4719	7754	2	1.7
LLDCs		*627*			*2.6*
World	2521	5901	8909	1.8	1.4

Source of data: UN (1998a).

influenced by the uncertain future of the independent state in a rapidly globalising world and the need for data of greater locational precision.

Most population growth in LDCs

Between 1950 and 1998 about 89% of world population growth took place in LDCs. By 1998 they accounted for four-fifths (80.0%) of the total world population, in comparison with only two-thirds (66.9%) in 1950 (Table 3.2). In contrast, the MDCs accounted for only one-fifth (20.0%) of world population in 1998, in comparison with roughly one-third (33.1%) at mid-century. The proportion of world population living in LDCs is certain to go on rising, because at the end of the century they accounted for about 97% of all population growth. So by 2050, the United Nations 1998 projections indicate that LDCs will contain 87% of the total world population, but of course by then the composition of the LDC category may have changed; some LDCs may have become MDCs, and vice versa. Already there is more of a spectrum than a dualism.

Nine of the top ten countries contributing to world population growth during 1995–2000 are current LDCs, the only MDC being the USA (Table 3.3). India and China together contributed 27.4 million per annum during that quinquennium, 35.3% of the world total growth, more than all the other eight countries combined. India now

Table 3.3. *The top ten annual contributors to world population growth, 1995–2000*

Country	Net addition p.a. (millions)	% of world pop. growth
India	16	20.6
China	11.41	14.7
Pakistan	4.05	5.2
Indonesia	2.93	3.8
Nigeria	2.51	3.2
USA	2.27	2.9
Brazil	2.15	2.8
Bangladesh	2.11	2.7
Mexico	1.55	2
Philippines	1.52	2
Top ten	**46.49**	**59.8**
World	**77.74**	**100**

Source of data: UN (1998a).

contributes 16 million or 21% of the world population increment (much more than China, with 11.4 million or 15%) and is projected to take over from China as the world's largest population.

In contrast, the populations of most MDCs, especially those in Europe, are actually declining and are expected to continue to do so, unless there is a huge increase in international migration from poorer to richer countries. For example, the 1998 United Nations projections suggest that the Russian Federation's population will fall massively by 26.2 million by 2050, Italy's by 16.2 million, Ukraine's by 11.6 million, Spain's by 9.4 million and Germany's by 8.8 million, in comparison with a decline of the UK's population by a mere 2 million.

Increasing demographic diversity of LDCs

In 1950, there was a fairly sharp demographic dualism between the two categories of MDCs and LDCs which matched the economic dualism, although neither category was homogeneous. While experiencing economic development during the nineteenth and early twentieth centuries, the MDCs had progressed to a greater or lesser extent and at differing rates through the various phases of demographic transition from high to low fertility and from high to low mortality, to reach a stage of population growth close to zero and ageing population structures. In contrast, the LDCs were mostly characterised by high fertility, high but generally lowering mortality and increasingly rapid population growth. Only in the last few decades has that situation changed as considerable declines in fertility began to slow down population growth in most LDCs. This occurred with or without accompanying economic development, although many of the poorest countries, often categorised as Least Developed Countries (LLDCs) and mostly found in sub-Saharan Africa, have experienced the least demographic transition (see Table 3.2).

Within the less developed world the demographic picture has diversified substantially since 1950. While many LDCs, particularly in East and South-east Asia (e.g. China, Malaysia, Thailand), have progressed rapidly through the demographic transition during a very short period, achieving low fertility, mortality and natural increase levels not unlike those of many MDCs, others, especially the LLDCs in sub-Saharan Africa, have progressed much less and many have been devastated by the toll of AIDS which has prevented gains in life expectancy. In consequence, the earlier demographic dualism of MDCs and LDCs, in terms of fertility and mortality levels at least, has been greatly attenuated. Indeed, while the range of levels of economic development between and within the world's countries has widened, that of rates of demographic change (fertility, mortality, natural increase) has narrowed, so that the common and persistent use of the crude economic dualism of MDCs and LDCs in describing demographic change is much less meaningful than it was. Even the World Bank's economic classification of low, middle and high income countries does not marry well with demographic change.

While economic contrasts should not be ignored as factors in demographic change, there is now a stronger case for considering the demographic impact of the contrasts between the world's main cultural realms or civilisations, nearly all of which incorporate hundreds of millions of people: Chinese, Hindu, Islamic, Western, East European,

Japanese, Latin American and African (Huntington 1996). These realms have had very different demographic histories and densities of human habitation which greatly influence their current population characteristics and dynamics. For example, patterns and dynamics of dense populations in the deeply rooted peasant civilisations of China and the Indian subcontinent were much less affected by European imperialism and civilisation than were the much sparser indigenous populations of the Americas, Africa and Oceania, including their patterns, levels and rates of urbanisation. There are also, for example, persistent important cultural contrasts in the roles and status of women and in attitudes to families that have great demographic significance. Diverse gender roles greatly influence population characteristics such as age of marriage, fertility rates, sex preference, the preponderance of male births, male/female differences in life expectancy and overall sex ratios (Clarke 2000). Although the recent global diffusions of major social changes, such as the small nuclear family system, the delayed marriage of women and the spread of female education, have penetrated all cultures, they meet greater or lesser resistance and are frequently regarded as undesirable imports from the West.

Unfortunately for demographers, these large cultural realms do not coincide neatly with the political boundaries of countries. While China, India and Japan are core countries of three of the realms, other realms do not have similar core countries, so demographic comparisons between them are less easy. Moreover, census data of countries often do not include data for ethnic, religious and cultural groups, especially those whose distributions overlap political boundaries and are seen as political threats. Thus, demographic data are not readily available for cultural realms, whose demographic distinctiveness, as for example between the declining populations of the former Soviet republics of eastern Europe and the still rapidly rising populations of the Islamic countries of western Asia, has been somewhat submerged by greater attention to geographical, economic and political groupings. Table 3.4 gives some indication of the contrasts in cultural realms through the diverse current average population ratios and growth rates of selected major populous countries. Obviously, the range would be much greater for smaller state populations, but their rates are often much affected by external migration.

Table 3.4. *Contrasts in large representative populations of different cultural realms, late 1990s*

Cultural realm & representative country	Pop. (millions) 1998	% Urban 1995	Total fertility rate	Life expectancy M/F	Av. pop. growth (% p.a.) 1995–2000
LDCs					
Chinese–China	1255.1	30	1.8	68.2/71.7	0.9
Hindu–India	975.8	27	3.07	62.1/62.7	1.6
Islamic–Pakistan	147.8	35	5.02	62.9/65.1	2.7
African–Nigeria	121.8	39	5.97	50.8/54.0	2.8
Latin American–Brazil	165.2	78	2.17	63.4/71.2	1.2
MDCs					
N. American–USA	273.8	76	1.96	73.4/80.1	0.8
W. European–UK	58.2	89	1.72	74.5/79.8	0.1
E. European–Russia	147.2	76	1.35	58.0/71.5	−0.3
Japanese–Japan	125.9	78	1.48	76.9/82.9	0.2

Source of data: UNFPA (1998: 110–18).

Changing continental balance in population distribution

Asia and Europe have long contained the largest concentrations of humanity, together incorporating as much as 88.3% of the world population in 1850. But with the spread of Europeans during the nineteenth century mainly to North and Latin America and Oceania, the subsequent rapid population growth within those continents, the slowing down of population growth in Europe, and a fairly stable situation in Asia and Africa, the continental balance of population distribution had changed markedly by 1950 (see Table 3.5). At that time, the proportion of the world population living in Asia and Europe had dropped to 77.4%, while the populations of the Americas and Oceania had risen from 5.1% in 1850 to 13.7%.

Since 1950 the continental balance has shifted strongly, largely as a result of contrasting rates of natural increase, as intercontinental

Table 3.5. *Percentage world population distribution by continental region, 1800–2050*

	1800	1850	1900	1950	2000	2050?
Asia	64.7	65.3	54.6	54.6	56.7	59.1
Europe & ex-USSR	21	23	26.9	22.8	16.2	7.1
Africa	10.9	6.7	9	8.9	13.1	19.8
North America	0.7	2.2	5.2	6.6	5.1	4.5
Latin America	2.5	2.7	4	6.6	8.4	9.1
Oceania	0.2	0.2	0.4	0.5	0.5	0.5

Source of data: United Nations Population Division, *World Population Prospects*, various.

migration has played only a minor role. Perhaps the most significant feature is the marked decline in the percentage living in Europe and the former Soviet Union, from 22.8% in 1950 to about 16.2% in 2000, lower than for many millennia. This decline will certainly continue, so that by 2050 Europe is expected to contain only 7.1% of the world population, less than a third of its percentage of the world population in 1950; surely this will have considerable implications for the global role of European culture. A lesser relative decline occurred in North America during 1950–2000 from 6.6% to 5.1%, and this seems likely to persist. In sharp contrast, substantial percentage increases have occurred in the rapidly rising populations of Africa and the less rapidly rising populations of Asia and Latin America, and these are projected to continue well into the twenty-first century. Inevitably, the poorer populations of the less developed world will account for an ever-increasing proportion of humanity, exacerbating some of the great regional inequalities.

So, today the New Worlds of the Americas and Oceania still only account for about 14% of the world's total population, 86% living in the Old World of Europe, Asia and Africa, and current projections suggest that those percentages will change little by 2050 (Table 3.5). Moreover, east and south central Asia still contain nearly half (49%) of humanity and the two mega-populations of China and India nearly 38%, but these proportions are likely to decline because the slow and potentially

Table 3.6. *Countries with more than 100 million inhabitants, 1998 and 2050*

Country	Population 1998 (millions)	Country	Population 2050 (millions)
China	1256	India	1529
India	982	China	1478
USA	274	USA	349
Indonesia	206	Pakistan	346
Brazil	166	Indonesia	312
Pakistan	148	Nigeria	244
Russia	147	Brazil	244
Japan	126	Bangladesh	213
Bangladesh	125	Ethiopia	170
Nigeria	106	Congo	160
		Mexico	147
		Philippines	131
		Vietnam	127
		Russia	122
		Iran	115
		Egypt	115
		Japan	105
		Turkey	101

Source of data: UN (1998a).

negative growth of the Chinese population will not be offset by much more rapid population growth in the Indian subcontinent.

In 1998 there were ten countries with more than 100 million inhabitants, including only three classified as MDCs (USA, Russian Federation and Japan), but by 2050 the United Nations projections suggest that there will 18, all the eight newcomers being current LDCs mostly in Africa or Asia: Ethiopia, Congo, Mexico, Philippines, Vietnam, Iran, Egypt and Turkey (Table 3.6).

Subtly changing boundaries of the inhabited area

The rapid growth of the world's population is concentrating in only a small part of the earth's inhabited area which itself occupies less than a third of the earth's surface and, despite massive world population growth, is remarkably stable. Known by Greek geographers as the ecumene, a term revived by German geographers in the nineteenth century, it largely coincides with the area of land being occupied by peasant societies for agricultural purposes.

The creation of Europes Overseas in so-called 'empty continents' during recent centuries expanded the ecumene considerably within the New World, but at much lower densities than in the long established peasant civilisations of the Old World. Physical restrictions of altitude, relief and climate rapidly imposed constraints upon extension of the inhabited area (Clarke 1972). The hostile environments of high latitude cold zones, hot and temperate deserts and high mountain ranges have largely ruled out human habitation over at least one-sixth of the earth's land area and curtailed it over more than a third, thus affecting habitation over more than half of the earth's surface (Noin 1997: 26).

Of course, these hostile environments were far from being uninhabited. Some have been the seats of great civilisations; some have been refuges. Consequently, many have been occupied by quite dense populations, often with considerable ingenuity. For example, the arid world (defined here by including the hyper-arid, arid and semi-arid zones) which covers 37.3% of the earth's surface has long contained an unusually large population, especially where abundant water supplies were available from external sources, such as the Nile, Tigris, Euphrates and Indus. Now it is variously estimated to contain about 15–19% of the world total population, a perhaps unexpectedly large percentage including many mega-cities (Clarke and Noin 1998).

The edges of the inhabited area are influenced by both human and environmental factors and are difficult to define, as much depends upon the scale of analysis and the threshold of population density that is regarded as an inhabited area. By the end of the nineteenth century the overall inhabited area was broadly established and, as in 1950, covered only about 30% of the earth's surface – it is difficult to be more precise – the rest being largely 'empty'. That percentage appears to have changed very little. Using 1994 population data assigned to 5 minute by 5 minute quadrilaterals covering the world, Tobler *et al.* (1995) found that just under 31% of the grid cells (of an average size of 54.7 sq. km.) were populated (Figure 3.2). Obviously, changing the grid size would alter the percentage, but not greatly. Noin (1997), comparing his world map of population distribution for 1990 with that of Söderlund (1960) for 1950–53, found a striking resemblance in the pattern of empty and populated areas, despite more than a doubling of population.

The reason why the overall pattern of the inhabited area persisted throughout the twentieth century is that in only a few parts of the world has recent rapid population growth led to the agricultural colonisation of formerly empty areas (e.g. Sumatra; the coastal islands of Bangladesh). Moreover, generally the sporadic and isolated occupation of empty areas by oil and mining companies and by tourist resorts has led to very little expansion of the broad pattern of world population distribution. Population concentration has been much more common in many parts of the world, with a general decline of population on the sparsely populated fringes of the inhabited area through the widespread demise of nomadic pastoralism and retreat from the more marginal agricultural lands (e.g. highlands, desert fringes). Rural depopulation of the margins has been most common in MDCs where the forces of globalisation have made them uneconomic and where out-migration and low natural increase or even natural decrease have meant that rural population pressures have declined. Evidence is abundant, as for example in many parts of Spain, France, Poland and Germany (Faus-Pujol and Higueras-Arnal 1986). In Britain, however, the juxtaposition of urban and rural, long-distance commuting and the acquisition of second homes have partially offset the process of rural depopulation.

Population redistribution and regional concentration

The second half of the twentieth century saw a huge reduction in the links between people and land. Only about 11% of the world's land surface is arable land, and it is generally believed that little extra land will be converted to cropland. With widespread declines in agricultural employment, huge increases in the proportion

Global Demography Project
NCGIA/Geography, UC Santa Barbara

Inh. per square km

5   10   25   50   100   250   500

Figure 3.2. World Map of Population Density, 1994, based on 5 minute by 5 minute quadrilaterals.
Source: Tobler *et al.* (1995).

employed in the secondary sector and particularly in the tertiary sector and in the proportion unemployed, the concept of an agriculturally defined ecumene becomes ever less meaningful. The populations of most countries are becoming less and less agricultural and rural, and by a combination of differential natural increase and increased migration they are concentrating more and more within their existing inhabited areas. This process of massive population redistribution is likely to continue at least until 2050.

Population redistribution and regional concentration have occurred in almost every country in the world, intensifying but not altering dramatically the overall pattern of the world population map. Densely peopled areas with high rates of natural increase along with economic core areas containing the major cities, industrial centres and ports have witnessed the most population growth, especially within unitary states that are highly centralised (e.g. Bangladesh, Colombia, Egypt, Iran, South Korea). Noin (1997: 18) calculated for 1990 that over three-fifths of the earth's population was concentrated on about one-tenth of its land area, with an average density of 230 per sq. km., six times the world mean. The proportion and the density are rising. In contrast, less successful agricultural or industrial areas, especially where natural increase is low, have experienced either limited population growth or even decline.

Although patterns of population distribution are not clearly differentiated according to levels of economic development, since 1950 they have been greatly affected by the wide differences in population dynamics. In many MDCs where there is currently both external and internal hypermobility (Adams 1999) but little or no natural increase, the effects of mobility upon overall patterns of settlement are not striking, because those patterns are well established; the population distribution is said to be in a state of dynamic equilibrium (Rowland 1979). While Australia's population doubled between 1945 and 1990 and movements in and out of the country multiplied a hundredfold (Haggett 1996), the pattern of peripheral population distribution remained broadly the same. Similarly, increased internal movement in Britain facilitated by an enormous increase in personal transport has not greatly altered the overall population map, except through counter-urbanisation, peri-urban developments and the resurgence of small and medium-sized towns. Adams (1999) has calculated that the average Briton travelled five miles a day in 1950 and about 30 miles today. It is doubtful that even the current huge growth in electronic communications will greatly alter the population maps of MDCs in the immediate future.

The situation is different in those LDCs where the patterns of settlement are not so long established and where strong core–periphery contrasts reflect previous patterns and processes of colonisation, as for example in much of sub-Saharan Africa, South America and South-east Asia. There, increasing population mobility accompanied by considerable natural increase has led to rapidly growing polarisation and spatial disequilibrium of population distribution. Efforts by governments to develop population redistribution policies (UN 1981; Brennan 1999) to reverse the process of population concentration in regions like south-east Brazil and Java have been singularly unsuccessful so far.

One aspect of this polarisation of population concentration in LDCs is the localisation of nearly two-thirds of the world's population living within 60 kilometres of the coast. Three-quarters of them live in tropical developing countries, often in densely peopled deltaic and low-lying areas that are greatly affected by floods and pollution and that are extremely vulnerable to the looming prospect of rising sea levels as a result of global warming (UNCED 1991). Moreover, the problems of coastal concentration are likely to be exacerbated, because the economic forces of increasing globalisation inevitably tend to augment the locational advantages of coastal situations and proximity.

Massive acceleration in the urban population

The pattern of population concentration is of course intensified by the massive acceleration in the numbers of people living in towns and cities. The speed of growth of the world's urban population since 1950 has been even more impressive than the growth of the total population from 2.5 to over 6 billion. In 1950, only 0.75 billion lived in towns, about 29.2% of world population, but the 1996 United

Nations projections of urbanisation (UN 1998c) suggest that by 2000 the number had nearly quadrupled to about 2.93 billion or 48.5% of world population. These projections reflect some recent slowing down of urban growth in LDCs and the exaggeration of earlier world urbanisation projections made by the United Nations in the 1970s (Brockerhoff 1999), but they indicate that by 2006 roughly half of the world population will be urban dwellers, and that by 2030 the total urban population will be of the order of 5.1 billions or three-fifths of the total world population.

It must be stressed that projections of urban populations convey less confidence than those of total populations, as they are based on inadequate urban data and are dependent upon varying national definitions of what is urban and what is rural, a very blurred dualism. Nevertheless, they give an order of magnitude to this immense change in world population distribution, suggesting that about 59 million new urban dwellers were being added annually in the late 1990s. Thus, three-quarters of the world's newcomers are living in towns and cities, and this fact reflects the enormous economic changes that have brought about huge changes in employment.

Perhaps even more striking is the estimate that 89% of the new urban dwellers live in LDCs, where between 1950 and 2000 the number of urban dwellers multiplied more than six times from 0.29 billion or 17% of the total LDC population to about 1.9 billion or 40% of the total. In contrast, the urban population of MDCs has grown much less rapidly from 0.45 billion (54% of the total) in 1950 to 1.04 billion (92%) in 2000. So, although the level of urbanisation in LDCs is less than half that in MDCs, their urban growth rate is currently much faster and they are catching up rapidly (Gilbert 1996). It is projected that by 2025–2030 LDCs will account for 98% of all new urban dwellers, and their recent urban growth would have been even greater if trends in productivity and terms of trade had not favoured agriculture more than manufacturing and thus slowed down the progress of rural–urban migration (Brockerhoff 1999). Even so, it was estimated that in 1999 three million rural dwellers in LDCs were still leaving their homes for the towns each year.

Of course, the level and rate of urbanisation varies greatly from one part of the world to another. It is fastest from the lowest levels in the poorer LLDCs of sub-Saharan Africa, where countries are often ill-equipped to cope with the problems associated with dramatic urban growth. As in Africa, the proportion of the Asian population living in cities is now just over a third, but the numbers involved are much larger and Asia is expected to have to absorb a total of 1.5 billion new urban inhabitants by 2030, over 0.55 billion in the Indian subcontinent alone (Brennan 1999). In Latin America and the Caribbean, urbanisation has proceeded so rapidly that the proportions urbanised are comparable with many of those in the more developed world, but the rates of urban population growth are much more rapid.

Unprecedented growth of mega-cities

In general, the density pattern of cities is highest where population density is highest, but city patterns are more complex because in many harsh environments (e.g. deserts) populations tend to concentrate even more than usual in large cities, which are accounting for an increasing proportion of the world's population and of the world's urban population. United Nations 1996 urbanisation projections indicate that by 2015 there will be as many as 523 cities with a million inhabitants or more (sometimes called 'millionaire cities'), containing 40% of the world urban population, compared with only 75 in 1950 containing 25%. However, cities with fewer than a million inhabitants, which contain the bulk of the world's urban population, are growing much more slowly (Brennan 1999).

A fundamental feature of urbanisation since 1950 has been the extraordinary emergence of mega-cities, formerly considered to be cities with 4 or 5 million inhabitants or more but now defined more commonly as those with 10 million inhabitants or more. Because of inadequate data upon which past projections were made and a deceleration in rates of LDC population growth, most of the United Nations 1970s projections of the growth of mega-cities proved to be very much too high by many millions, especially those for Mexico City, São Paulo, Shanghai, Jakarta, Seoul and Cairo. However, those for Lagos and Dhaka were much too low, because during 1970–90 these two cities grew very rapidly, by 6.7% and 7.6% per annum respectively (Brockerhoff 1999).

Despite the lowering of projections, mega-city growth during the second half of the twentieth century was still phenomenal and un-precedented, particularly because it was largely located within LDCs, including some of the poorer ones. In 1950, New York was the only city in the world with more than 10 million inhabitants, but by 2000 there were 14, led by Tokyo but including 10 located within LDCs. The 1996 projections of urbanisation look forward cautiously to 2015 when there may well be 26 mega-cities, 22 of them in LDCs, and containing 418 million inhabitants or 10.6% of the world urban pop-ulation (UN 1998c). Table 3.7 suggests that in 2015 the world's largest city will still be Tokyo but the next seven largest cities will be Mumbai (Bombay), Lagos, São Paulo, Dhaka, Karachi, Mexico City and Shanghai, all major cities of populous LDCs that have more than 100 million inhabitants. So all of these rapidly growing mega-cities are expected to be much larger than New York.

Of course, there is no simple explanation of the rapid growth of mega-cities, but its relationship to national population growth rates and numbers is quite remarkable (Brennan 1999). Thus, few mega-cities are found within MDCs where slow city growth (e.g. Tokyo, Osaka, New York) or 'urban exodus' (e.g. London) prevails. Nearly three-quarters of those projected to reach mega-city status by 2015 are located within the top ten countries contributing to world population growth (see Table 3.3). Undoubtedly, this accounts for much of the projected growth of the 16 mega-cities projected for 2015 that are located in Asia; seven of these are in the Indian subcontinent alone, and all are projected to grow much more rapidly than, for example, the mega-cities of Latin America where national population growth rates have slowed down much more. Not surprisingly, as the future growth of mega-cities seems to be heavily influenced by national population growth rates, many feel that more fertility control would have a greater effect on limiting their surging populations (if not their areal dimensions) and on facilitating their governance than would various attempts to restrain in-migration which, as in Jakarta and Manila, have so far failed fairly ignominiously.

Table 3.7. *Projected growth of mega-cities, 1995–2015*

Mega-city	Pop. 1995 (millions)	Est. pop. 2015 (millions)	Est. % growth p.a. 1995–2015
Tokyo, Japan	27	28.9	0.3
Mexico City, Mexico	16.6	19.1	0.7
São Paulo, Brazil	16.5	20.3	1
New York, USA	16.3	17.6	0.4
Mumbai (Bombay), India	15.1	26.3	2.8
Shanghai, China	13.6	18	1.4
Los Angeles, USA	12.4	14.2	0.7
Calcutta, India	11.9	17.3	1.9
Buenos Aires, Argentina	11.8	13.8	0.8
Seoul, Rep. of Korea	11.6	13	0.6
Beijing, China	11.3	15.6	1.6
Osaka, Japan	10.6	10.6	0
Lagos, Nigeria	10.3	24.6	4.5
Rio de Janeiro, Brazil	10.2	11.9	0.8
Delhi, India	9.9	16.9	2.7
Karachi, Pakistan	9.7	19.4	3.5
Cairo, Egypt	9.7	14.4	2
Tianjin, China	9.4	13.5	1.8
Manila, Philippines	9.3	14.7	2.3
Jakarta, Indonesia	8.6	13.9	2.4
Dhaka, Bangladesh	8.5	19.5	4.2
Istanbul, Turkey	7.9	12.3	2.2
Tehran, Iran	6.8	10.3	2.1
Hyderabad, India	5.5	10.5	3.3
Lahore, India	5	10	3.5
Hangzhou, China	4.2	11.4	5.1

Source of data: UN (1998c).

With the exception of Mexico City and Beijing, mega-cities are all coastally located, and not surprisingly more than two-thirds of those expected to reach this status in 2015 will also be on the coastal fringes. Although they comprise only a small percentage of the total world urban population, their continued growth will further intensify the increasing concentration of population in these ecologically vulnerable zones.

Growing environmental impacts of mega-cities

The massive process of population concentration and the mercurial emergence of mega-cities exemplifies the reduction of environmental controls over population distribution within the inhabited area, and humanity's increasing detachment from land. Furthermore, urban centres and mega-cities in particular cause many environmental problems. It is perhaps surprising that recent environmental concerns seem to have focused more on preserving and conserving the world's open spaces than on attending to the environmental problems of population concentration and the plight of the milling masses, despite the interrelationships of environmental concerns. Attention has been addressed more to human impact upon the world's forests, oceans, ice caps, deserts, wildernesses and biodiversity than to the impact of mega-cities on their local environments and their inhabitants: the declining and contaminated water supplies, the accelerating atmospheric pollution, the severely inadequate sanitation facilities and the enormous quantities of solid and liquid waste for disposal.

Environmental stress is variously experienced by different mega-cities. For instance, the most rapidly growing mega-cities of the poorer and less economically dynamic countries of south central Asia and Africa (see Table 3.7) tend to face especially severe problems differing from those of other continental regions, and certainly from the very much slower growing mega-cities of wealthier countries like the USA and Japan that are much more able to tackle the problems. The speed of population growth and levels of poverty in mega-cities such as Lagos, Dhaka and Karachi pose immense infrastructural problems (e.g. power, housing, sanitation, roads, hospitals) for their administrations, especially in coping with the many forms of pollution. It has been said dramatically that many such mega-cities are being inundated by their own chemical and faecal wastes (Brennan 1999), but they are also choking on their own emissions of sulphur dioxide and lead. Somewhat ironically, in Rio de Janeiro, the beautiful city of the

1992 Earth Summit, there were large demonstrations in March 2000 against the collapse of its sewerage system and the effects of oil pollution and soaring sewage levels on its renowned beaches and its lakes. Inevitably, the poorest inhabitants of the poorest mega-cities living in the worst housing conditions suffer the most environmental stress. Closest to the foci of pollution, they tend to experience the highest rates of unemployment, under-employment, malnutrition, morbidity and mortality. In consequence, the long-standing presumption that living conditions are better in larger cities than in the countryside is only true where efficient city management and governance occur. Unfortunately, they are lacking in many LDCs, especially in sub-Saharan Africa where 'mega-villages' of several hundred thousand people have emerged (Brockerhoff and Brennan 1998).

Although the rates of growth of the mega-cities of LDCs are uncertain, at least in the short-term future there can be little doubt that they will continue to grow and to localise millions of people living in poverty. They have become part of a global network of capital, trade, communications and crime, linking them more and more with each other and distinguishing them from the countries in which they are situated. Consequently, the many internal inequalities of LDCs are intensified.

Conclusion

With the world's growing population concentrating on less and less land and linking together more and more tightly in an expanding network of large cities within 'the global village', the well-established geographical concepts of inhabited area, state populations and population density are gradually becoming less meaningful and relevant. The physical environment imposes itself strongly on the sparsely peopled and uninhabited areas, which largely remain unattractive to human occupancy and are often depopulating, so that they are sometimes called 'negative'. It imposes itself much less on the world's inhabited areas, where the rapidly changing human geography responds especially to the very diverse patterns of population growth and economic growth, including the growing impact of globalisation, by ever-increasing population concentration in economic core areas of countries, notably within the burgeoning cities of LDCs.

References

Adams, J. (1999). *The Social Implications of Hypermobility*, The Environment Directorate. Paris: Organisation for Economic Cooperation and Development.

Brennan, E.M. (1999). Population, urbanisation, environment, and security: a summary of the issues. *The Woodrow Wilson Center, Environmental Change and Security Project Report*, Issue 5, 4–14.

Brockerhoff, M. (1999). Urban growth in developing countries: a review of projections and predictions. *Population and Development Review*, **25**, 757–778.

Brockerhoff, M. and Brennan, E.M. (1998). The poverty of cities in developing regions. *Population and Development Review*, **24**, 75–114.

Clarke, J.I. (1971). Milling masses and open spaces. In *Biological Aspects of Demography*, ed. W. Brass, pp. 31–55. London: Taylor & Francis.

Clarke, J.I. (1972). Geographical influences upon the size, distribution, and growth of human populations. In *The Structure of Human Populations*, ed. G.A. Harrison and A.J. Boyce, pp. 17–31. Oxford: Clarendon Press.

Clarke, J.I. (2000). *The Human Dichotomy. The Changing Numbers of Males and Females.* Oxford: Elsevier Science (Penguin).

Clarke, J.I. and Noin, D. (1998). *Population and Environment in Arid Regions.* Man and the Biosphere Series Vol. 19. Paris: Parthenon Publishing for UNESCO.

Faus-Pujol, M.C. and Higueras-Arnal, A. (1986). *Rural Depopulation.* IGU Commission on Population Geography, Zaragoza, 25–30 August 1986. Zaragoza, Spain: Department of Geography and Spatial Organization.

Gilbert, A. (1996). Third world urbanization. In *Companion Encyclopedia of Geography. The Environment and Humankind*, ed. I. Douglas, R. Huggett and M. Robinson, pp. 391–407. London: Routledge.

Haggett, P. (1996). Geographical futures. Some personal speculations. In *Companion Encyclopedia of Geography. The Environment and Humankind*, ed. I. Douglas, R. Huggett and M. Robinson, pp. 965–973. London: Routledge.

Huntington, S. (1996). *The Clash of Civilisations and the Remaking of the World Order.* New York: Simon & Schuster.

Noin, D. (1997). *People on Earth.* World Population Map. Paris: UNESCO.

Rowland, D.T. (1979). *Internal Migration in Australia.* Australian Bureau of Statistics. Canberra: Census Monograph Series.

Söderlund, A. (1960). *Jordens Befolking* [Peopling of the Earth]. Stockholm: Stockholm School of Economics.

Tobler, W., Deichmann, U., Gottsegen, J. and Maloy, K. (1995). *The Global Demography Project*. Technical Report 95-6. University of California, Santa Barbara: National Center for Geographic Information and Analysis.

UN (1981). *Population Distribution Policies in Development Planning*. Population Studies No. 75. New York: Department of International Economic and Social Affairs.

UN (1998a). *World Population Prospects: The 1998 Revision*. New York: UN Population Division, Department of Economic and Social Affairs.

UN (1998b). *World Population Projections to 2150*. New York: UN Population Division, Department of Economic and Social Affairs.

UN (1998c). *World Urbanization Prospects: The 1996 Revision*. New York: UN Population Division, Department of Economic and Social Affairs.

United Nations Conference on Environment and Development [UNCED] (1991). Relationship between demographic pressures, unsustainable consumption patterns, development and environmental degradation. Doc. A/CONF. 151/PC/46.

UNFPA (1998). *The State of World Population*, The New Generations. New York: UN Family Planning Association.

4

Population, community and society in peasant societies

ROBERT LAYTON

Definitions

In contrast to the two very similar definitions of population provided in the two previous chapters by a demographer and by a geographer, in evolutionary biology a population is viewed as a construct involving a set of organisms that share a genetic ancestry and interbreed. A *community*, on the other hand, is a set of people with common interests, values and distinctive patterns of social interaction, often occupying a single geographical area. *Society* consists of all the people encompassed by a total, relatively self-sufficient network of institutions within which their social interactions take place. Peasant communities have been described as 'part societies' (Redfield 1960). A *peasant* is a cultivator whose productive activities are primarily directed toward his/her household's subsistence needs, but who is under some degree of economic and political obligation to powerful outsiders and/or carries out subsidiary production for a market operating in the wider society. Peasants also interact on a daily basis with other members of their local community.

Through an examination of the interaction between *population*, *community* and *society* in peasant communities, this chapter will illustrate how studies of peasants can throw light on the relationship between the sociocultural environment and the dynamics of human population biology affecting individuals' survival and reproductive success, i.e. mortality and fertility. The chapter begins by summarising a long-running debate concerning the relative importance of the individual and the ecological or social system to which that individual belongs. It will then look at the transition from hunting and gathering to peasant agriculture, illustrating continuities in human social behaviour that may have a genetic basis, but also showing how the new mode of subsistence created a new natural and sociocultural environment and assessing its effects. The chapter will then look at the impact of customary rules of inheritance on household structure and inter-household relations in European peasant communities, in order to show how culture influences the individual's opportunities for biological reproduction.

The intrinsic properties of the individual and the emergent properties of interaction

The difference between a population and a community or society is fundamental to social anthropology. It is axiomatic in social anthropology that social behaviour cannot be completely explained in terms of the intrinsic properties of individuals. This view was forcefully expressed by the French sociologist Durkheim at the start of the twentieth century. Durkheim argued that

- During social interaction (co-operation, reciprocity, competition, domination), social forces are created which either limit or enable our ability to achieve our personal goals.
- The language we speak, the currency with which we buy and sell goods and the legal system that guides our social behaviour, are the product of specific cultural traditions.

(see Durkheim 1938 [first published 1901].)

The Durkheimian tradition has recently been challenged by the evolutionary psychologists, Tooby and Cosmides. They argue that evolution has given the human mind a complex structure with many specific skills which enable the individual to learn a language and to decide with whom to mate, when to co-operate with others, and how much parental care to expend on each offspring (Cosmides *et al.* 1992: 73). These skills developed through natural selection during the long period in which our species lived by hunting and gathering, and are virtually universal among members of our species. Society and culture derive directly from the structure of the mind.

The issue is to a large extent one of research focus. For Tooby and Cosmides the fact that some people speak English and others do not is trivial when compared with the fact that 'everyone passes through a life-stage when the same species-typical language acquisition device is activated' (Cosmides *et al.* 1992: 45). Social anthropologists such as Geertz, on the other hand, are interested in culturally specific 'webs of significance' (e.g. Geertz 1973a,b).

The argument is not new. Durkheim engaged in a lengthy debate with his contemporary Tarde over the relative importance of the individual or the social system in explaining how the Industrial Revolution had come about. Tarde regarded the inventor as an individual with a special psychology, who combined existing ideas or practices in a novel yet logical fashion (see papers republished in Tarde 1969). Durkheim rejected appeals to the 'psychology' of the inventor. He argued that social change emerges from the interaction of communities and is a collective process. Just as animal populations adapted to competition by finding specialised ecological niches, so human communities responded by adapting the economy to specialise in wine making or cereal cultivation, coal mining or steel production (Durkheim 1933: 266–7). Durkheim therefore proposed an ecological theory of social change.

In the latter half of the twentieth century a number of writers have followed Tarde's approach to social change, comparing the spread of innovations to the spread of a virus (Cavalli-Sforza 1971; Cloak 1975; Dawkins 1976; Cavalli-Sforza and Feldman 1981). Durham (1991) and Boyd and Richerson (1985) have looked critically at this idea. Does the 'epidemic model' adequately represent the transformation of peasant communities after the Industrial Revolution? Scott (1976), for example, showed how the 'Green Revolution', which introduced new high-yielding varieties of rice to Asian farmers, exacerbated divisions between the relatively prosperous and poorer peasants. Often, only the relatively wealthy could afford the chemical fertilisers and hire the labour or the simple machinery needed to cultivate 'Green Revolution' crops successfully. While the 'Green Revolution' varieties gave a better yield in the long run, if properly cultivated, they might fail once every five years. A wealthy peasant might have sufficient financial resources to survive, but the greater long-term yield would be of no use to a peasant household forced out of production by star-vation in the fifth year. Susceptibility to innovation is not randomly distributed. If a few are 'infected' with an innovation, this may make it harder for others to adopt it. Durham objects that the epidemic model is based on 'radical individualism' and generally ignores power and coercion, which are emergent properties of social interaction. 'In cul-tural systems . . . significant evolutionary forces can and do arise from unequal social relations' (Durham 1991: 182–3).

A parallel debate is in progress among evolutionary theorists as to whether the primary motor of evolution is the gene, or the ecological system that exerts selective pressures on the organism. Dawkins (1976) advocates the former, although recognising that genes interact with one another within the developing organism (Dawkins 1976: 271). Kauffman (1993) and Conway Morris (1998) argue that the envi-ronment to which organisms adapt is transformed by interaction. Kauffman writes, in a similar vein to Durham, 'In co-evolutionary processes, the fitness of one organism or species depends upon the characteristics of the other organisms or species with which it inter-acts, while all simultaneously adapt and change' (Kauffman 1993: 33). The position of Tooby and Cosmides is closer to that of Dawkins (see Cosmides *et al.* 1992), whereas Durkheim emphasised what today would be called the self-organising properties of social systems which emerge through human interaction. In this view, the advan-tages to the individual of genetically governed traits, such as the abil-ity to co-operate and to detect those who cheat during co-operation, emerge out of social interaction. Genes and their social environment 'co-evolve': each progressively changes the other. Both must be taken into account.

Genes and society in the transition between foraging and peasant farming

Examples of the co-evolution of genes and culture that followed the domestication of plants and animals are well known. Malaria-carrying mosquitoes such as *Anopheles gambiae* require warm, sunlit pools of fresh water in which to breed. This habitat is rare in tropical forest, but greatly increased by slash-and-burn cultivation. Sufficient selective pressure to increase the incidence of the sickle-cell trait to modern frequencies would have existed only after the introduction of agriculture. Populations such as hunter-gatherers that have no domestic livestock to milk display low frequencies of the gene which enables lactose absorption in adults. Two subgroups of dairy traditions that drink large quantities of fresh milk, East African pastoralists and northern Europeans, have been subject to genetic selection in favour of the ability to digest lactose in adults. Others, such as the North African and Mediterranean dairying peoples, have developed cultural techniques for processing milk into soured or fermented products which break down the lactose (Durham 1991; Holden and Mace 1997). In both cases, learned cultural practices have transformed the environment in which natural selection takes place.

The transition from hunting and gathering to farming transformed the adaptive niche of human populations. There are, however, continuities in social behaviour that provide evidence in support of the hypothesis that certain human social skills are genetically determined, and have persisted from the time when humans evolved as hunter-gatherers. Dunbar (1993) found that in different primate species the relative size of the part of the brain called the neocortex correlates with the size of social communities in each species. He deduced that humans were adapted to living in groups of between 150 and 200 individuals. A hunter-gatherer 'tribe' consists of a set of bands who speak the same dialect and regularly interact. A number of studies suggest that the regional community of interacting bands is of about the same size as a typical peasant village, that is, between 150 and 500 people (Kelly 1995 cf. Dunbar 1993). Hunter-gatherers move between bands where they have friends and relatives to exploit temporary local abundances of food and to escape drought. In a peasant village, the entire community is living in a single place. The geographical span of social relations declines, but the number remains constant.

All humans, including both hunter-gatherers and peasants, have the ability to keep track of whether partners in exchange are honouring reciprocal social obligations. Trivers (1985) has argued that this is also a genetically determined skill which has helped humans adapt to uncertain environments. Peasant farmers, like hunter-gatherers, are subject to unpredictable or stochastic variability in their environment that creates risk (Winterhalder 1990). Like hunter-gatherers, peasants minimise risk both through their subsistence strategies and through social relationships between households in the community. Winterhalder compares the peasant strategy of dispersing fields and diversifying crops to the hunter-gatherer strategy of relying on more than one prey species in a patchy environment. However, while humans have a unique capacity to maintain reciprocal social relationships, hunter-gatherers and peasants use this ability in different ways. Hunter-gatherers reduce risk by sharing the meat obtained from hunting large game animals and moving between bands to avoid local scarcity.

Population growth

There are other important differences in the environments of hunter-gatherers and of farmers that emerge from the practice of agriculture. Most, if not all, hunter-gatherers suffer seasonal shortages of food which determine the sustainable population density. The Gunwinggu of northern Australia told Altman (1987) that the three wet-season months were the hardest, until foraging could be supplemented with food purchased at the local government settlement. If women had worked at gathering for 63 hours per week during the wet season, the highest productivity they could have achieved would have been 1,600–1,800 kilocalories per day. It is likely, then, that hunter-gatherer populations were limited by annual seasons of shortage. It is generally agreed that women's body fat levels are also crucial to the regulation of population size among hunter-gatherers. Wilmsen (1989) compared foraging and pastoral !Kung in the Kalahari, and found the seasonal weight loss of women living by foraging was double that of pastoral women. Forager women average 4.5 live births, pastoral and sedentary !Kung average seven live births. Jones (1980) estimates that birth-spacing among the Gidjingali was halved, from 4–5 to 2–2.5 years between surviving children, when the Gidjingali were settled at Maningreda by the Australian Government. The conditions on government or mission communities in the twentieth century were unlikely to resemble those in early farming communities, and it is unlikely that farming produced a comparable, immediate population explosion. Body fat levels may still limit ovulation in a traditional peasant community (Panter-Brick *et al.* 1993). Nevertheless, farming did result in population growth, and peasant societies are subject to longer-term demographic cycles than have been demonstrated in hunter-gatherer populations.

In his 1803 *Essay on the Principle of Population*, Malthus argued that any population has the potential to increase faster than its food supply (see Hinde, Chapter 2). Diseases of overcrowding and famine are inevitable. The peasant population of Europe, up till the nineteenth century, did indeed exhibit the kind of cycles Malthus predicted. The French historian Braudel (1990) has since identified a series of cycles, each lasting several centuries, signalled by rise and fall in population numbers and changes in the rate of economic activity. Celtic, Roman, Merovingian and Carolingian Gaul each correspond to a population cycle.

Inheritance

While hunter-gatherers inherit from their parents rights to forage in particular areas, peasant farmers have developed more specific cultural rules for the transmission of rights to cultivated land. The study of peasant households has shown how inheritance rules and recruitment of labour reveal intimate connections between reproduction and subsistence production and culture. In western Europe there are two widespread rules of inheritance: unigeniture and partible inheritance. In unigeniture, the entire holding passes to a single heir. In partible inheritance the property is divided between all children. The two rules have radically different effects on the structure of social relations. Partible inheritance tends to level out differences of wealth between households, but unigeniture creates a landowning élite. Under unigeniture non-inheriting brothers and sisters may remain at home reduced to the role of servant, seek work as agricultural labourers or migrate to town. Partible inheritance is practised today in southern

Germany, north-east France and Switzerland, while unigeniture is practised in northern Germany, south-west France and Austria.

Since both occur in the Alps, the two inheritance systems cannot in any simple way be explained as adaptations to different physical environments. Inheritance rules do, however, have close links with reproductive strategies. They are determined by two conflicting goals: conserving a holding which is adequate to perpetuate the family line, and giving all children the best possible start in life (Cole and Wolf 1974). Unigeniture gives priority to the first, partible inheritance gives priority to the second. Augustins (1990) argued that partible inheritance and unigeniture have fundamentally different social objectives. Unigeniture aims to perpetuate the house (*maison*, *casa* or *ostal*) associated with a family line at the expense of disinherited kin. Partible inheritance aims to perpetuate a wider kindred even if it entails sacrificing the continuing association between a family line and its ancestral home (Augustins 1990, cf. Barthelemy 1988). Partible inheritance is associated with a higher rate of village endogamy (marriage within the village), because husband and wife can only use all the land they inherited if this lies within the boundary of one village community.

Marriage and inheritance strategies may also provide culturally specific ways of reducing population growth rates. Malthus travelled through Europe looking for evidence of the cycles he had predicted. Despite finding widespread evidence, he found an exception in the Alps, where the population seemed to be stable. He concluded that families in small, isolated mountain villages could see they were dependent on a limited supply of land. Malthus argued that the peak demand for labour occurred at harvest time, to obtain sufficient hay to feed the cattle while they were in their stables during the winter. The number of cattle was limited by the supply of hay. However, the size of the labour force available to harvest the hay was limited by the food supply (meat and cheese from the cattle). The amount of hay that could be grown was in turn determined by supply of manure, which depended on the number of cows. Each element in the subsistence regime limited the others. In order to avoid starvation, men and women delayed marriage to hold the number of children born to the carrying capacity of the land.

Alpine peasants never quite achieved the degree of rational planning implied in Malthus's model. In fact, a peasant to whom Malthus spoke explained that, even though late marriage was needed to prevent over-population and bring birth and death rates into equilibrium, he himself had married young (Malthus 1973 [1803]). Viazzo (1989) studied historical population patterns in the Alps and concluded that late marriage and adult celibacy did slow down population increase, but never quite enough to stabilise the population. None the less, after epidemics, age at marriage fell, and when the population had recovered age at marriage rose again. Three centuries of demographic data encompassing ten mortality crises in the Swiss village of Törbel showed that, while some emigration was necessary to offset population growth, fertility rose after a period of high mortality and declined during periods of population growth (Netting 1981). Smith (1981) argues that in England between 1541 and 1871, marriage rates rose or fell in response to real wage levels in agriculture, but with a time lag of about 30 years. As real wages rose, servants (who were typically unmarried) were tempted to delay marriage to take advantage of the higher income they could earn.

McGuire and Netting (1982) found that more children survived to the age of 20 in wealthy Törbel families, but a higher proportion were obliged to remain unmarried, to prevent fragmentation of the family holding. The poorest villagers, if they married, had fewer surviving children, but more of their children married. Thanks to partible inheritance, however, very few wealthy families in Törbel succeeded in maintaining their position over several generations.

Data from European village studies indicate that partible inheritance often tends to be associated with strategies among landowners to restrict birth rates and hence prevent increasing division of family holdings. This constraint does not apply in communities practising unigeniture. In the French village of Chanzeaux, some farms are rented and some owned by those who work them. Rented farms cannot be divided; they are transmitted as a single tenancy. Forty-five per cent of the population on rented farms was under 21 years of age, whereas on the smallholdings subject to partibility only 29% of the population was under 21, indicating that smallholding families were restricting their birth rate (Wylie 1966). Partible inheritance

also encourages adult celibacy. There are, for example, higher proportions of childless couples and unmarried adults living with a married
brother in the north Italian village of Tret, where partible inheritance is practised, than in the neighbouring village of St Felix, which
practises unigeniture (see Cole and Wolf 1974).

The household, production and inter-household relationships in the community

Differential access to labour and land are the two determinants of social inequality in a peasant community. According to Segalen (1991),
a farm of 13 hectares in the Breton commune of Bigouden required
a permanent work-force of 11 persons. Wylie (1966) estimates that
in 1841 a 30-hectare farm in Chanzeaux (in the Vendée, south of
Brittany) might require a work-force of four, with additional labour
at harvest time. Labour requirements clearly vary according to the
agricultural regime, although authors disagree about exactly how
they vary (see Flandrin 1979; Viazzo 1989; Sabean 1990; Segalen
1991).

The more active adults there are in a household, the greater the
household's output. Partible inheritance, on the other hand, makes
large families a threat to the future of the exploitation. A successful household operating within a regime of partibility will maximise
the available labour but minimise the number of heirs by delaying
children's marriage and encouraging adult celibacy. In parts of Tibet
where partible inheritance has also been the rule, peasant households practised polyandry, one wife marrying a set of brothers. Since
the woman's childbearing capacity is the critical element limiting the
number of heirs (see Hinde, Chapter 2), the effect on the household's
reproductive output is the same as if one brother had married and
the others remained celibate (Durham 1991; Crook 1995).

Laslett (1972) argued that nuclear families have always been the
norm in European peasant households. Flandrin (1979) rightly replies
that if one wants to understand how social differentiation takes place
in peasant communities, it is households on the margins, rather than
the average, that need to be investigated (cf. Scott 1976; Durham

1991). Although one of Laslett's English village censuses from 1676 reveals that although the average size of households was 4.47 persons, over half the population of the village was living in the households of the twelve men who owned almost all the land, provided employment for other families and were the political leaders of the village (Flandrin 1979).

Demographic factors alone can drastically limit the frequency and duration of complex family households where there are members of three generations or more than one married couple living together. Laslett (1972) recognised that, even if older people expect to live with their married child(ren) and grandchildren, a three-generation family will be present only for part of the household cycle, since the members of the older generation will almost certainly die during their grandchildren's childhood. Berkner's (1972) analysis of an eighteenth-century census from Lower Austria showed that male peasants married when they were about 25. If a son who would survive to adulthood were born within three years, this son would be ready to take over the farm when his father was in his late fifties. The father would be dead before the grandchildren reached adulthood. Sixty per cent of Berkner's households headed by a man aged 18–27 contained members of three generations, whereas this was true of only 5% of those headed by a man aged 48–57. The household labour force will ebb and flow with each generation, unless workers are hired in.

Panter-Brick (1993), following Erasmus (1956, 1965), identifies several ways in which a peasant household can recruit labour, and assesses the relative cost of each source. Networks of reciprocal, non-contractual assistance are integral to life in peasant communities throughout the world. They constitute what Scott called 'the moral economy of the peasant' (Scott 1976). Panter-Brick has shown how the degree of participation in mutual aid networks has a measurable effect on the health of Nepalese peasant women (Panter-Brick *et al.* 1993). De Waal (1989) showed that Sudanese peasants were more concerned about their inability to discharge reciprocal obligations during famine, which threatened the survival of mutual aid networks, than they were about going hungry.

The more active adults there are in the household, and the fewer non-productive, elderly or very young people, the greater the household's output, as long as it has enough land to work. The smaller the work-force in the household, the lower is the output. Single parent families with young children will be particularly vulnerable. A free-holding household will keep more of its output than a tenant household which has to pay rent, or provide a share of produce or corvée labour to its landlord. Working in the Italian Alps prior to the introduction of tractors, Cole and Wolf (1974) found that larger holdings in both the villages they studied sometimes hired agricultural labourers, while medium-sized ones managed through reciprocal labour. The smallest landholders were the most likely to hire themselves out to others (Cole and Wolf 1974 cf. Erasmus 1956).

Partible inheritance encourages village endogamy (see above). This in turn promotes mutual aid between households related by kinship, whereas in areas characterised by unigeniture, households prefer to be self-sufficient, or to rely on hired labour. Where unigeniture and patronage prevail, anthropologists working in Europe describe an unequal pattern of exchange. Rogers (1991) found that labour flowed disproportionately from small to large holdings in Aveyron, while farm equipment was loaned in the opposite direction. In St Felix, it was the large landowners that organised 'hay-mowing bees' attended by all the young men of the village, who were rewarded with food, drink and a small wage (Cole and Wolf 1974: 173–4). Wylie (1966) found that mutual aid in Chanzeaux was tainted by the suspicion of patronage favouring one, larger farmer.

External relations

Peasant agriculture can produce surplus that a central power can exploit by demanding tribute or taxes. This has no parallel within hunter-gatherer societies, except among the Calusa of Florida as reconstructed by Marquardt (1988), although hunter-gatherer economies have been distorted by demands for tribute from colonial powers, for example the Saami, described by Mulk and Bayliss-Smith (1999). Unlike hunter-gatherers, however, peasant communities are by definition not isolates but, in Redfield's phrase, 'part-societies', embedded in a centralised polity which charges tax or tribute, regulates trade and legislates on village society (cf. Wolf 1982). This is exemplified clearly below in the chapter by Kunstadter (Chapter 9).

The history of unigeniture in northern Europe illustrates the effect of power differences in the wider society on peasant communities. In England and Germany, unigeniture was imposed on populations that previously practised partibility. Anglo-Saxon England was characterised by partible inheritance. From the thirteenth century English feudal lords required fiefs to be transmitted undivided, thus imposing unigeniture on their tenants (Faith 1966; Flandrin 1979; Goody 1983). If villeins divided up the land among their heirs, progressively smaller holdings would increasingly cater only for their tenants' subsistence and yield lower profits. 'Primogeniture, the system most favourable to seigniorial interests, developed, probably under seigniorial pressure, where lordship was strong and where demesne farming became important' (Faith 1966: 85).

Unigeniture in the German-speaking Italian Tyrol was established early during colonisation, beginning in the eleventh century. At that time feudal lords agreed to hereditary tenure of tenancies in exchange for tenants' undertaking not to divide their holdings between heirs (Cole and Wolf 1974), and this can be compared with Viazzo's (1989) study in the Austrian Alps. This explains why the German-speaking village of St Felix practises unigeniture, while the neighbouring Italian-speaking village of Tret does not. In Upper Swabia, a region of southern Germany whose ecology is suited to pasture, grain production and forestry, lords took steps to enforce unigeniture throughout the region during the sixteenth century. In most of nearby Württemburg, however, wine production required intensive production and considerable risk, favouring densely populated villages and small units of production (Sabean 1990). Tenants here were allowed to continue practising partible inheritance.

Peasants' vulnerability to exploitation by the powerful illustrates dramatically how their life chances are determined by the distribution of power in the social system, to which they contribute through trade, and tribute or taxation. The effects of inheritance rules in Europe are mild in comparison with the plight of peasant communities in the

Third World. This chapter therefore ends with the example of the Guatemalan peasants studied by Wilson (1991). In the early 1980s, many Maya worked as bonded serfs on plantations and cattle ranches. During this period, anti-government guerrillas carved out a power base in the area. Their proposed reform of land ownership was attractive and many local people joined them. The government responded by attacking the civilian population on whom the guerrillas depended for support, burning villages, destroying crops and massacring people. The guerrillas were unable to defend the villagers and encouraged them to flee to the forest. By 1983 the main guerrilla army had been defeated and withdrew; the Mayan villagers fell under government/army control between 1983 and 1987 as hunger and lack of protection forced them out of the forest. The army resettled the Maya in camps near towns or plantations, or on army bases. They were 're-educated' with films extolling freedom and prosperity in the United States, while US evangelical groups were allowed to proselytise. At the time Wilson was writing, army-supported vigilante patrols held political power in the villages, and the army settled disputes (cf. Werbner 1991 on the impact of colonisation on indigenous farmers in Zimbabwe).

Conclusion

Peasant communities are often about the same size as hunter-gatherer 'tribes'. Both peasants and hunter-gatherers depend heavily on reciprocal social relationships to reduce risk. The common thread in social organisation may well derive from the genetic evolution of social skills. Peasants, however, have different residence patterns and labour requirements from hunter-gatherers. Farming transformed the ecosystem and the diet of peasants, and modified demographic processes. Reliance on cultivated fields demands new systems of inheritance, which in turn reshapes relations within and between households. Social interaction generates the social niche within which individuals' life chances are constrained. While cognitive skills may well be genetically inherited, the environment they must cope with has been transformed by the emergent properties of social interaction and culture change, both within peasant communities and in the wider social systems to which they belong. These changes generate new patterns of social interaction and new cultural procedures that in turn reshape the dynamics of human population biology.

References

Altman, J.C. (1987). *Hunter-Gatherers Today. An Aboriginal Economy in North Australia.* Canberra: Aboriginal Studies Press.

Augustins, G. (1990). *Comment se Perpetuer? Devenir des Lignées et Destins des Patrimoines dans les Paysanneries Européennes.* Nanterre: Société d'Ethnologie.

Barthelemy, T. (1988). Les modes de transmission du patrimoine: synthèse des travaux effectués depuis quinze ans par les ethnologues de la France. *Études Rurales,* 110–112: 195–212.

Berkner, L.K. (1972). The stem family and the developmental cycle of the peasant household: an eighteenth-century Austrian example. *American Historical Review,* **77**, 398–418.

Boyd, R. and Richerson, P.J. (1985). *Culture and the Evolutionary Process.* Chicago: University of Chicago Press.

Braudel, F. (1990). *The Identity of France,* Volume II: *People and Production* [trans. S. Reynolds]. Glasgow: Collins.

Cavalli-Sforza, L.L. (1971). Similarities and differences in sociocultural and biological evolution. In *Mathematics in the Archaeological and Historical Sciences,* ed. F.R. Hodson, D.G. Kendall and P. Tautu, pp. 535–541. Edinburgh: Edinburgh University Press.

Cavalli-Sforza, L.L. and Feldman, M.W. (1981). *Cultural Transmission and Evolution: A Quantitative Approach.* Princeton, NJ: Princeton University Press.

Cloak, F.J. (1975). Is a cultural ethology possible? *Human Ecology,* **3**, 161–182.

Cole, J.W. and Wolf, E.R. (1974). *The Hidden Frontier: Ecology and Ethnicity in an Alpine Valley.* New York: Academic Press.

Conway Morris, S. (1998). *The Crucible of Creation. The Burgess Shale and the Rise of Animals.* Oxford: Oxford University Press.

Cosmides, L., Tooby, J. and Barkow, J. (1992). Introduction: evolutionary psychology and conceptual integration. In *The Adapted Mind: Evolutionary Psychology and the Generation of Culture,* ed. J.H. Barkow, L. Cosmides and J. Tooby, pp. 4–136. Oxford: Oxford University Press.

Crook, J.H. (1995). Psychological processes in cultural and genetic coevolution. In *Survival and Religion: Biological Evolution and Cultural Change,* ed. E. Jones and V. Reynolds, pp. 45–110. London: Wiley.

Dawkins, R. (1976). *The Selfish Gene*. Oxford: Oxford University Press [page reference is to the 1989 edition].

de Waal, A. (1989). *Famine that Kills*. Oxford: Clarendon Press.

Dunbar, R. (1993). Coevolution of neocortical size, group size and language in humans. *Behavioural and Brain Sciences Evolution*, **16**, 681–735.

Durham, W.H. (1991). *Co-evolution: Genes, Culture and Human Diversity*. Stanford: Stanford University Press.

Durkheim, E. (1933). *The Division of Labour in Society* [trans. G. Simpson]. London: Macmillan [French edition 1893].

Durkheim, E. (1938). *The Rules of Sociological Method* [trans. S.A. Solovay and J.H. Mueller]. London: Macmillan [French edition 1901].

Erasmus, C.J. (1956). Culture, structure and process: the occurrence and disappearance of reciprocal farm labour. *Southwestern Journal of Anthropology*, **12**, 444–469.

Erasmus, C.J. (1965). The occurrence and disappearance of reciprocal farm labour in Latin America. In *Contemporary Cultures and Societies of Latin America*, ed. D.B. Heath and R.N. Adams, pp. 173–199. New York: Random House.

Faith, R.J. (1966). Peasant families and inheritance in medieval England. *Agricultural History Review*, **14**, 77–95.

Flandrin, J.L. (1979). *Families in Former Times: Kinship, Household and Sexuality* [trans. R. Southern]. Cambridge: Cambridge University Press.

Geertz, C. (1973a). Thick description: towards an interpretive theory of culture. In *The Interpretation of Cultures*, ed. C. Geertz, pp. 3–30. London: Hutchinson.

Geertz, C. (1973b). Deep play: notes on the Balinese cockfight. In *The Interpretation of Cultures*, ed. C. Geertz, pp. 412–453. London: Hutchinson.

Goody, J. (1983). *The Development of Family and Marriage in Europe*. Cambridge: Cambridge University Press.

Holden, C. and Mace, R. (1997). Phylogenetic analysis of the evolution of lactose digestion in adults. *Human Biology*, **69**, 605–628.

Jones, R. (1980). Hunters in the Australian coastal savannah. In *Human Ecology in Savannah Environments*, ed. D. Harris, pp. 107–146. London: Academic Press.

Kauffman, S. (1993). *The Origins of Order: Self-organisation and Selection in Evolution*. Oxford: Oxford University Press.

Kelly, R.L. (1995). *The Foraging Spectrum: Diversity in Hunter-Gatherer Lifeways*. Washington, DC: Smithsonian Institution Press.

Laslett, P. (1972). Introduction: the history of the family. In *Household and Family in Past Time*, ed. P. Laslett and R. Wall, pp. 1–73. Cambridge: Cambridge University Press.

Malthus, T. (1973 [1798, 1803]). *An Essay on the Principle of Population*, Books I and II. London: Dent.

Marquardt, W. (1988). Politics and production among the Calusa of south Florida. In *Hunters and Gatherers: History, Evolution and Social Change*, ed. T. Ingold, D. Riches and J. Woodburn, pp. 161–188. Oxford: Berg.

McGuire, R. and Netting, R. McC. (1982). Leveling peasants? The maintenance of equality in a Swiss Alpine community. *American Ethnologist*, **9**, 269–290.

Mulk, I.M. and Bayliss-Smith, T. (1999). The representation of Sámi cultural identity in the cultural landscapes of northern Sweden. In *The Archaeology and Anthropology of Landscape*, ed. P. Ucko and R. Layton, pp. 358–396. London: Routledge.

Netting, R. McC. (1981). *Balancing on an Alp: Ecological Change and Continuity in a Swiss Mountain Community*. Cambridge: Cambridge University Press.

Panter-Brick, C. (1993). Seasonal organisation of work patterns. In *Seasonality and Human Ecology*, ed. S.J. Ulijaszek and S.S. Strickland, pp. 220–234. Cambridge: Cambridge University Press.

Panter-Brick, C., Lotstein, D.S. and Ellison, P.T. (1993). Seasonality of reproductive function and weight loss in rural Nepali women. *Human Reproduction*, **8**, 684–690.

Redfield, R. (1960). *The Little Community/Peasant Society and Culture*. Chicago: University of Chicago Press.

Rogers, S.C. (1991). *Shaping Modern Times in Rural France: the Transformation and Reproduction of an Aveyronnais Community*. Princeton, NJ: Princeton University Press.

Sabean, D.W. (1990). *Property, Production and Family in Neckarhausen, 1700–1870*. Cambridge: Cambridge University Press.

Scott, J.C. (1976). *The Moral Economy of the Peasant: Rebellion and Subsistence in Southeast Asia*. New Haven, CT: Yale University Press.

Segalen, M. (1991). *Fifteen Generations of Bretons: Kinship and Society in Lower Brittany, 1720–1980* [trans. J.A. Underwood]. Cambridge: Cambridge University Press.

Smith, R.M. (1981). Fertility, economy, and household formation in England over three centuries. *Population and Development Review*, **7**, 595–622.

Tarde, G. (1969). *On Communication and Social Influence* [trans. T.N. Clark]. Chicago: University of Chicago Press.

Trivers, R. (1985). *Social Evolution*. Menlo Park, CA: Benjamin/Cummins.

Viazzo, P.P. (1989). *Upland Communities*. Cambridge: Cambridge University Press.

Werbner, R. (1991). *Tears of the Dead: the Social Biography of an African Family*. London: Edinburgh University Press.

Wilmsen, E.N. (1989). *Land Filled with Flies: a Political Economy of the Kalahari*. Chicago: Chicago University Press.

Wilson, R. (1991). Machine guns and mountain spirits. *Critique of Anthropology*, **11**, 33–61.

Winterhalder, B. (1990). Open field, common pot: harvest variability and risk avoidance in agricultural and foraging societies. In *Risk and Uncertainty in Tribal and Peasant Economies*, ed. E. Cashdan, pp. 67–87. Boulder, CO: Westview Press.

Wolf, E. (1982). *Europe and the People without History*. Berkeley: University of California Press.

Wylie, L. (1966). *Chanzeaux, a Village in Anjou*. Cambridge, MA: Harvard University Press.

5

From genetic variation to population dynamics: insights into the biological understanding of humans

JAUME BERTRANPETIT AND FRANCESC CALAFELL

The extent of human biological variation

Humans are biological organisms. After chapters on demographic and socioeconomic perspectives on human populations and their spatial distribution, it is now appropriate to focus on contemporary ways in which world-wide human biological diversity and dynamics can be studied through human population genetics in its development in the era of the Human Genome Project.

The description of human differences is ancient. There are references in antiquity to humans with different physical features, such as those described by Herodotus, but the confusion between biological features and those we would today consider cultural was common until the twentieth century and still persists in some material. Over the course of the last century, there have been continuous debates about human diversity, but these debates were dominated by descriptions of morphological characteristics, for example pigmentation, and body shape and size. These descriptions in turn gave rise to 'racial' classifications, which remain highly significant socially. Even with the increase in understanding of genetic processes, much of this body of work has done nothing to further the understanding of human diversity, because in most cases the analyses did not take into consideration two basic issues.

The first issue is the evolutionary perspective on a given pattern of diversity. Following a tradition in the description of nature itself, the description of human populations continued to take, even recently, a naturalistic perspective, more in accordance with a pre-Darwinian view of detailed description than with its meaning in biological terms. This kind of description has been one of the more vain and fruitless developments of biology, related more to a collector's enthusiasm than to scientific endeavour. This perspective not only embedded the morphological descriptions of humans, but even affected the descriptions of gene variant (allele) frequencies, the study of which was for a long time no more ambitious in scope. Many of the 'new data' provided by the frequencies of hundreds of genetic markers were given in the literature just for the descriptive interest itself. When 'racial classifications' began to incorporate genetic data, their scope did not alter except in the use of more 'modern' phenotypic data. Now when we view a given pattern of diversity in given, defined populations and in a given moment in time, we see it as the result of evolutionary dynamics.

The second issue relates to the genetic bases of the differences being described. Most of the traditional morphological descriptions of human populations used traits of dubious genetic significance. Let us take as an example the long-lasting devotion to measuring skulls in order to describe human variation. One of these measures, the cephalic index, after decades of being considered one of the main descriptors of human diversity, is assumed today to be uninteresting because of the extent of the non-genetic factors affecting it and the uncertain genetic contribution to it. Ignorance of the genetic basis of a given biological character precludes the possibilities of interpreting its significance in population dynamics, because evolution works on the genetic constitution (gene pool) of all populations and species. Although the genetic basis should be taken into consideration, this does not mean that any clear genetic character must be identified; in fact most of the genetic variants in humans have become informative only when they have been used to answer specific questions beyond the initial descriptive purpose.

Taking the genetic information as the starting point for description enables the use of the whole theoretical framework provided by molecular evolution and population genetics, seldom possible when a morphological trait is the basis for consideration. The relationship between the genetic character and the phenotype of the individual of course needs careful consideration, since many genetic changes will have no expression, and thus will be selectively neutral, while others will cause complex interactions with other genetic and non-genetic factors in the most inclusive definition of the 'environment'. Therefore, an important way of looking at human populations biologically is to consider the genetic diversity within and between populations. One can describe the genetic structure of populations by estimating the frequencies of different alleles. The recognition that these frequencies are the result of dynamic processes makes those processes an essential topic in this volume.

Variation at the molecular level

The Human Genome Project, an international programme to sequence the whole genome and map all human genes, and other initiatives are producing a vast knowledge of our genome, including the variation that exists between different individuals. Genes are widely understood to be segments of DNA, which are themselves made up of strands of four basic chemicals known as nucleotides. Most of human genetic variation has now been pinpointed to the level of alterations in these single nucleotides, otherwise known as single nucleotide polymorphisms (SNP). The number of specific sites where more than one nucleotide has been described is in the order of millions. Chromosomes are DNA strands, which carry the genetic material inherited equally from each parent; in each normal human cell there are 46 chromosomes. Of relevance to points in this chapter and in Chapter 8 (Jorde *et al.*) is that these 46 chromosomes are 22 pairs of non-sex chromosomes, known as autosomes, and two sex chromosomes. In females there is a pair of X chromosomes and in males an unmatched pair of an X chromosome and a small Y chromosome. Other DNA occurs independently of the chromosomes: this is called mitochondrial DNA (mtDNA); as mtDNA from sperm cells does not enter the egg cell in fertilisation, the only mtDNA passed on to sons and daughters is that from the mother. Y chromosomes can only be passed from fathers to sons, and mtDNA can only be inherited along the maternal line, whereas all other chromosomes come in pairs, one copy from the father and one from the mother.

A diploid set of chromosomes means a set of pairs, while a haploid set means just one of each pair originating from one or other parent.

Several aspects are of interest here.

1 *The biological meaning of SNP variation.*

Most SNP variation is neutral and is not affected by natural selection. This is clearly the case for at least 98% of SNPs, as they are located in genome regions that do not code for proteins. The functional meaning of nucleotide differences between humans is, therefore, very limited. Among those in coding regions, 25% are synonymous, that is, code for exactly the same protein. The rest, just three-quarters of 2%, may have different functional implications.

2 *The amount of variation in humans.*

The extent of sequence variation can be measured in a very simple way. For a given nucleotide position in a population sample, the statistical measure known as nucleotide diversity would range from 0 if all individuals carried the same nucleotide at that position to 1 if all were different. Usually, nucleotide diversities are measured as averages over a stretch of DNA sequence. The observed nucleotide diversities in different genome regions in humans show a range of values from 0.00005 to 0.0192, almost a 400-fold spread. However, most of the variation is related directly to specific properties of the genomic regions. For instance, nucleotide diversity in Y chromosome sequences is rather low (from 0.00005 to 0.0004) probably just because of the population effective size of the Y chromosome. For every copy of Y chromosome in a population, there are four copies of each autosome (non-sex chromosome). Thus, and just because of numbers, autosomes enjoy roughly four times the chance of accumulating variation than the Y chromosome (Pérez-Lezaun *et al.* 1997). Nucleotide diversity in autosomes ranges from 0.0005 to 0.0020, with one exception, the HLA system, which jumps to 0.0192. In this region, which contains the HLA genes involved in recognising 'self' versus 'non-self' and in the recognition of pathogens, natural selection has favoured diversity as a way of fighting a wider array of pathogens.

3 *The (geographic) structure of the variation.*

In addition to different levels of variation distributed across genome regions, genetic variation is also clustered in different sets of individuals. The largest differences between human groups are between men and women. As stated above, the Y chromosome is found only in men, and it contains approximately 30 million nucleotides that are not found in women. Over the 3,000 million nucleotides in the total haploid set per individual, this is a 1% difference, roughly the same amount as the difference between humans and chimpanzees! However, the Y chromosome is mostly 'junk' DNA and rather poor in genes: just a dozen seem to be sufficient to steer foetal growth to the development of a male.

The distribution of genetic variation among individuals can be analysed hierarchically, by grouping individuals into populations and populations into large continental groups. The results of this kind of analysis show that most of the genetic variation is found between individuals rather than between populations or between continental groups. On average, 85% of the genetic variance can be explained by differences between individuals belonging to the same population, 7% by differences between populations of the same continent and 8% by differences between continental groups (Barbujani *et al.* 1997). This has social implications that deserve a wide dissemination. It may just be a consequence of a pattern discussed further below that human populations, however defined, have a shorter history than the genes they carry.

A general view of the genome variation may be obtained from rough measures of difference. Considering that the average difference between individual genomes is one per thousand, for a haploid genome (half of our diploid genome or what we received from one of our two parents, containing around three thousand million nucleotides) there will be three million differences. If 2% are in coding regions, and 75% of the substitutions are non-synonymous, it may represent around 45,000 expressed point differences, a non-negligible number that may account more for differences between individuals than for differences between populations, as discussed above.

None the less, when considering the amount of genetic diversity in other species, including apes, it becomes evident that the genetic diversity among humans is small. That is mainly due to our relatively short history as a species, which means a lack of time for the accumulation of a greater divergence between populations and continental groups. The differences that affect the phenotypic characteristics that are evident in the external differences observed in the normal range of human variation have not yet been identified at the molecular level, although they must exist.

Variation and divergence: what makes a human human?

The extent of variation among humans is small, even if the number of genetic differences is in the order of tens of thousands. How does this observation fit with the differences between humans and our closest relatives, the chimpanzees?

Humans and chimpanzees are different in a number of morphological and cognitive aspects; the latter are still debated, and it is beyond the scope of this chapter to define precisely what makes us human. However, we should like to stress that what makes humans human is not conceptually different from what makes mice mice. That is, the human cognitive features may be regarded as nothing but a set of adaptive traits. The nascent field of evolutionary psychology is producing quite reasonable hypotheses: for instance, on the adaptive value of complex language and context-dependent reasoning. Therefore, the human specificity is neither more distinguished nor more special than the mouse specificity, and most of the reasoning that follows could be applied to the adaptive differences of any one species compared with its closest kin.

For many years, evolutionary biology has stressed our close relationship to apes, basically using a genetic concept. Although the morphological and behavioural differences between humans and chimpanzees seem obvious, in genetic terms the difference is in the order of 1% of our genetic information. In fact, recent studies of DNA sequences show that the similarity is even higher than 99%. To be puzzled by 'only' a 1% difference is highly subjective. For a haploid genome that apparently small fraction may represent 30 million differences in nucleotides and even if the great majority are in non-coding regions, there is ample room for substitutions having profound effects on the specific biology of humans.

At the present day, our ignorance of the basic genetic differences that could account for the human specificities is great. A report (Gagneux and Varki 2001) concludes that there is only one clearly known fixed difference between the two species with change of function. It is clear that small structural differences in genes related to development may cause large phenotypic changes, but their identification is elusive. Moreover, it is likely that, besides basic functional changes, other changes will have accumulated independently after the split between the lineages leading to the two species. These changes would be difficult to interpret, as their phenotypic effect may not be related to the adaptive effects that drove human evolution. Large differences in biology may be attributable to a small number of genetic differences. Meanwhile, most of the genetic substitutions do not have any functional consequence and are carried by the individuals without influencing their fitness. These are genome regions that have been affected by historical factors through the generations and provide a good indication of the demographic past.

Genetics and the evolutionary process

Genetic diversity exists as a result of an evolutionary process and should be understood within a framework of population dynamics which genetic analysis may help to unravel.

For many years now, genetics has been at the centre of evolutionary thought, even if the modern synthesis of evolution was achieved in a much wider arena, in which palaeontology, the natural sciences and genetics met with the Darwinian ideas of evolution by natural selection. Much has been expected of genetics (and still is). It is clear that medical genetics and genomics have stirred the field with promises about health, and have been able to attract investment to achieve the Human Genome Project, with its detailed description of our genome. What has not been so clearly explained is that the deeper knowledge of our genome is nothing more than a golden gate to a future understanding of the intimate fabric of life and to new possibilities of understanding life, from individuals to populations to species, in their uniqueness and in their commonality.

A detailed knowledge of our genome will also allow determination of the extent of genome variation, which is a key to the comprehension of the evolutionary process illuminated by two different but fully linked aspects. On the one hand, there are the evolutionary dynamics of the genome and its parts, chromosomes, genes and polymorphisms, and on the other the dynamics of populations, the reproductive units within which the genome is transmitted from individuals to others in the following generation.

These are completely intertwined processes which have been, until recently, considered separately. Evolutionary genome dynamics was long considered basic molecular genetics, since it deals with biochemical processes, such as mutation, and with biological constraints, such as the cellular architecture of chromosomes. However, it has opened its interpretation, and now it embraces issues that have consequences beyond genetic change. For example, the new analyses of genetic variation from the standpoint of evolutionary genome dynamics allow recognition of the process by which changes accumulate in each gene or genome fragment. That is, they allow reconstruction of a gene tree where tempo and mode of change are detailed. This kind of analysis leads also to uncovering the 'footprint' of selection; through analysis of molecular diversity, there can even be recognition of past positive (advantageous) or negative (purifying) selection.

On the other hand, population genetics has a tradition of relying on the basic factors that change allele frequencies in populations, namely mutation, selection, drift and migration (see below). The interpretation of genetic diversity from a population genetics perspective led to inferences on demographic population history: it allowed hypotheses on population expansions, founder effects and admixture, often in conjunction with hypotheses picked out from other disciplines such as palaeoanthropology, archaeology and linguistics.

These two ways of interpreting genetic diversity remained in tight isolation until recently. The same patterns of genetic diversity could be analysed by rather similar techniques, but could receive widely different interpretations from a genetic anthropologist who would declare, for instance, that some population expansion has been traced, and from a geneticist usually involved in studying fruit fly (*Drosophila*) populations and perhaps keen on tracing selection.

Fortunately, those days seem to be over and we are seeing the fusion of two separate disciplines, population genetics and molecular evolution. Genetic diversity is now open to new questions, which unlock attractive possibilities of future interdisciplinary development. The two disciplines together can deal with the understanding of the mechanisms producing variation (whose consequence is evolution), by understanding the genetic process and the dynamics of the populations where the genetic processes took place. This conjunction provides the tools with which to reconstruct the whole evolutionary process.

Impact of evolutionary forces

Evolution is the process of change in the genetic makeup of a population over time. Consequently, the most basic component of the evolutionary process is the change in gene frequencies (Graur and Li 2000). This is why population genetics has focused on the understanding of evolution through the effect of evolutionary forces. An unsolved problem is the extent to which the comprehension of the forces at the molecular level fully explains the tempo and mode of the production of the diversity of life. When dealing with a single and young species, humans in this case, the small-scale process and the low degree of morphological differentiation make the findings at the molecular level much easier to interpret. Moreover, as theoretical models can be carefully tested, there are plenty of possibilities for learning the dynamics of genes, which in turn allow us to infer past events in populations.

The main forces that shape the genetic structure of populations are mutation, selection, drift and migration. Mathematical models exist for each of them and their interactions. These models have become a key factor in recognising the relative importance of each process for the current, observed frequency of a given gene or genome region, and so allow the estimation of the population parameters involved.

While the basic dynamics and consequences of each of these processes are well known, there has been much debate on the relative importance of random drift and adaptation-related selection. As will be seen, drift emerges as an important force, able to explain not only unusual cases but also the general dynamics of genetic change.

Mutation

Mutation represents changes in DNA sequences. It has long been considered as the raw material for evolution, as all the innovations must have appeared first through the mutation process. There is no doubt that this is the case, even if mutation has a connotation more related to the production of disease than to the production of benign innovation. This perception has a clear biological base. Proteins are the basic units of all organic function. They are made up of amino acids, coded for by DNA. A change in the nucleotide sequence of DNA may change just one amino acid in the unique protein synthesised. When, in the very complex system of any living organism one of its parts changes, it is highly likely that it will produce a dysfunction in the whole system. It is very unlikely that its function will improve, which is why mutations frequently cause genetic disease.

Mutation occurs at very low frequency. It is low enough for the likely dysfunction it will produce not to harm maintenance of the population, but high enough for benign novelties eventually to spread and allow evolution. The perception of mutation as a 'necessary harm' does not do justice to the innovative force it represents from the perspective of the population and is instead an anthropocentric view resulting from concentration on the importance of the individual. Medical genetics, of course, pays attention to the relationship between genes and disease, since the discipline's primary aim is to understand and treat genetic diseases. It is proving its own success. The evolutionary view, however, goes beyond the individual and explains the existence of genetic disease as a necessity for the future.

Selection

Fertility and mortality are topics that are central to population dynamics and to this volume. In the modern synthesis of Darwinian ideas and genetic information, selection is deemed to occur following the differential fertility and/or mortality of certain individuals due to the presence or absence of one or more of their genes. Several key issues are crucial to understanding selection:

1 *How has selection shaped human evolution?*
To answer this the effect of selection must be unravelled from that of other confounding factors. Once the role of selection has been established, the importance of the several selective forces should be assessed, namely:
  (i) *Negative or purifying selection*, where the frequency of deleterious mutations decreases; this is the clear case of almost all mutations that produce disease, mainly in Mendelian inherited disorders.
 (ii) *Positive selection*, where advantageous mutations are being selected for.
(iii) *Balancing selection*, where an equilibrium is reached at intermediate frequencies, usually through some kind of advantage of the heterozygotes over both homozygotes, or through frequency dependent selection.

2 *How can one measure the strength of selection acting on the frequency of characteristics in a group?*
Where selection is the causal explanation for adaptation, it may be measured in general terms for a species, for a population or for a given gene or genomic region.

There is no doubt that negative selection has been, and still is, important in human evolution. As mutations mostly produce deleterious variants, these are eliminated from the population by purifying selection. The number of these variants is extremely high and today several web pages keep track of the disease-related mutations in an increasing number of genes (for a good starting point, see the Human Gene Mutation Database at http://archive.uwcm.ac.uk/uwcm/mg/hgmd0.html). Purifying selection has also been called normalising natural selection, as it keeps the gene pool constant in response to a constant environment.

Meanwhile, positive selection has long been assumed to be the main cause of adaptation. Among the variants produced by mutation, even if most are deleterious, some may give a better adaptation to a given environment, improving a pre-existing phenotype, thereby increasing its frequency in the population by the action of selection. Although this scenario may seem common sense, in humans it has been proved to have occurred in only a few genes, none of which is of major functional importance. In fact, positive selection may be recognised at the molecular level as a dearth of genetic diversity near the genetic variant being selected for. As the new, enhanced phenotype grows in frequency in the population, the genetic background around the new variant also increases in frequency. It is not by chance that this phenomenon is called a 'genetic sweep' or, from the point of view of the genetic background, 'genetic hitchhiking'. If the new variant reaches fixation in the population (a point reached faster if the advantage of the new variant is greater), so does the genetic background around it, and all pre-existing variation in that particular genomic region would be lost. At that point, the molecular clock is reset to zero and variation starts accumulating again slowly by mutation. The result is a genomic region that is impoverished in variation compared with other regions of the genome. That is a telltale sign of positive selection.

Views on selection have changed dramatically in the last few decades. Having moved away from a 'selectionist' viewpoint, in which explanations emphasise the effect of the positive and balancing modes of selection as the main driving forces in the evolutionary process, the field has shifted towards a 'neutralist' viewpoint, which stresses the effects of mutation, random genetic drift (see below) and purifying selection (Graur and Li 2000). Both points of view can easily be formulated in terms of models that predict variation at the molecular level. Given that predictions differ between models, it is possible to test them empirically. When the analysis of selection has reached the molecular level, it has been possible to analyse with accuracy the effect of selection. Most results conformed to the neutralist model, with the result that the neutralist view has become well established in the field.

This is the reason why some of the discussions on selection are now seen as old-fashioned, including the splendid essay by Dobzhansky (1972) entitled 'Natural selection in mankind'.

Genetic drift

Sometimes called random genetic drift, this is the force that produces random fluctuations in allele frequency following the random sampling of gametes in the process of reproduction. Sampling occurs because only a very small number of the gametes will produce individuals in the next generation and there will be, by chance, variation. In the same way that deviations from a 50:50 ratio will be larger when tossing a coin a small number of times than if it is tossed a very large number of times, this variation is more likely to be important if population size is small. This again emphasises the interrelationship with demographic variables.

It is easy to understand the effects of genetic drift in specific circumstances, such as the founder effect (Diamond and Rotter 1987). This refers to the establishment of a new population by a few original founders, who as migrants carry only a small fraction of the total genetic variation of their parental population. There are specific cases where its effect has been demonstrated through the high frequency of a genetic disease carried within a population that expanded from a few original settlers. Several diseases are known to have a high frequency among such groups, for example the Afrikaners of South Africa, the French-speaking population of Québec and the Finnish population. In all these cases there is a 'specific genetic heritage' due to a founder effect. Although more difficult to detect, drift must have played an important role in the differentiation of human populations in ancient times when population densities and mobility were low. In fact, drift must be the main mechanism that explains genetic differentiation among human populations for neutral markers.

At the molecular level, drift seems to have had a central role in evolution, because it has become evident (Kimura 1983) that the majority of molecular changes in evolution are attributable to the random fixation of neutral (or nearly neutral) mutations. In summary, stemming from the molecular analysis of the genome, drift is now considered a main driving force in the evolutionary theory.

Migration

As referred to in earlier chapters, the demographic variables of migration and mobility are notoriously difficult concepts to define with precision. Short-range movement of individuals is one of the factors that make the definition of 'populations' so difficult. When migrants move from one population to another this clearly has demographic and socioeconomic effects on the donor and recipient populations. As long as those individuals reproduce in their new population, the movement of individuals between populations also entails the movement of genes. The genetic consequences of migration are a function of two variables: the amount of difference in the gene pools of both populations (usually measured as the *genetic distance* between them), and the relative demographic contribution of the newcomers to the recipient population. It should be noted that a constant, small trickle of immigrants over many generations can have the same effect as a sudden movement of a larger population. Also, repeated short-range mobility by generations of individuals can, over time, cause long-distance migration of genes.

Through DNA studies, the genetic analysis of migration can be framed in terms of admixture: a mixed population can be identified, as well as the populations that probably contributed to the mixture. In certain cases, the hypothesis of admixture itself can be tested; in others, admixture is taken for granted and the analysis seeks to estimate the contribution of each parental population. Recent examples of the latter type of analysis are those carried out on Nubians (Krings *et al.* 1999) and Brazilians (Alves-Silva *et al.* 2000; Carvalho-Silva *et al.* 2001). As discussed below, the accuracy of this kind of analysis depends on the genealogy of the gene used in the analysis. This genealogy is usually more ancient than the ethnogenesis process, that is, the set of social and demographic processes that led to the formation of a given population. Admixture analyses based on population-specific lineages generated by an ancient gene genealogy are much more precise than those that employ only the frequencies of ubiquitous alleles.

Two genome regions, mitochondrial DNA (mtDNA) along the maternal line and the Y chromosome along the paternal line, provide excellent tools for the analysis of sex-specific migration patterns. If there were differences in mobility between the two sexes, the genome region transmitted exclusively by the more mobile sex would tend to be more homogenous across populations. This was shown to be the case for mtDNA as compared to the Y chromosome on a global scale (Seielstad *et al.* 1998). The conclusion from that observation was that virilocal marriages, a practice prescribed in many cultures, meant that women spread their genes across populations much more than men did. However, it was pointed out (Stoneking 1998) that that study was based on different sets of samples for the two markers, which is not an optimal research design, whereas the studies by Jorde *et al.* (Chapter 8) on sex-specific social mobility were based on markers from identical individuals.

An analysis of four central Asian populations (Comas *et al.* 1998; Pérez-Lezaun *et al.* 1999; Calafell *et al.* 2000) addressed the unravelling of sex-specific migration patterns from genetic data. They studied four population samples, two from highland populations, the Kazakh and the highland Kirghiz, both living above 2,000 m, and two from lowland populations, the Uighur and the lowland Kirghiz, and they typed the mtDNA sequences and the Y-chromosome markers of the same individuals. They described within-population genetic diversity and genetic differentiation between populations, as well as the origin of mtDNA lineages from elsewhere. They found that, in all four central Asian populations, roughly one-third of mtDNA lineages had their origins in west Eurasia, while the remaining fraction was shared with east Asian populations. Thus, they concluded that the central Asian mtDNA sequence pool had originated from two distinct, already differentiated sets of lineages, found at the western and eastern edges of the Eurasian continent.

The mtDNA and Y-chromosome lineages in the above samples from central Asia showed different diversity patterns. The mtDNA sequences were highly diverse within the populations, both lowland and highland, but the four populations were not genetically different from each other for mtDNA. The paternal lineages (that is the DNA on the Y chromosomes) showed quite a different pattern: lowland populations were internally much more diverse than highland populations, and genetic variation was clearly compartmentalised between populations. The joint interpretation of the two diversity patterns points to two demographic processes: the founding of the highland populations by a reduced number of individuals, and higher female mobility. It is known that the highland habitats in central Asia were first settled just a few centuries ago; thus, a founder effect on genetic diversity is still traceable. The individuals that first colonised the highland habitats carried with them a subsample of the genetic diversity found in their populations of origin and, to this day, genetic diversity in the highlanders is still reduced. However, that is so only for the Y chromosome while, as stated above, the mtDNA sequences were highly diverse within each population but quite homogeneous across populations. The most likely explanation for this phenomenon is a higher female mobility, possibly related to virilocal residence, that would have spread mtDNA sequences across populations, even to the point of replenishing mtDNA diversity in highland populations.

This is but an example of how the interpretation of genetic diversity in current populations can throw light on demographic practices rooted in cultural patterns.

Genome versus population understanding

A purely biological endeavour such as the analysis of diversity in our genome has implications for a wide variety of disciplines, since the pattern of diversity depends not only on the internal dynamics of the genome but also on the demographic, social and geographical history of the population. Population genetics deals not only with mutation and selection but also with social and spatial mobility and migration, including differential sexual and selective patterns of movement. The classical volume by Cavalli-Sforza and Bodmer (1971) is still a key reference.

The main factors affecting the gene pool are:

1 *Genetic mechanisms*, the processes of basic biological production of variation in the genome, independent of individual-related considerations, such as mating patterns, population structure or stratification. Some of the genetic mechanisms are intrinsic to the genome, like mutation or recombination, while others are environment-dependent, like natural selection, which encompasses a wide variety of forms.

2 *Mechanisms dependent on population history, culture and demography*: these involve processes dependent on population size and its change through time, on population structures (geographical, sociocultural, hierarchical, sex-related, etc.) and on mating behaviour (preferences, prescriptions, proscriptions and inter- and intra-group mating patterns). All these act on the genome through population size, expansion events or migration. Population size affects the extent of genetic drift. Expansion events tend to reset the molecular clock, and most of the current diversity in an expanded population has its roots at the onset of the expansion. Migration blends the genetic pools of the admixing populations. In summary, the demographic and social factors most involved in genetic change are population size (founder effects and bottlenecks being the best known), population expansions, the social structure of the population, including subdivisions, migration (including admixture) and mating behaviour (polygyny, polyandry, assortative mating, endogamy, consanguinity, etc.).

Some genetic mechanisms produce a 'footprint' in the genome that may mimic the effect of a demographic event, but analytical tools have been devised which can recognise most of the past events. For example, lack of genetic diversity produced by selection (either negative or positive) may mimic the effect of a population expansion, where a small number of founders had caused lack of diversity. The analysis of several genes may, however, clarify whether a reduced diversity was due to selection, since this would have affected different genes in broadly different ways given their different biological functions, or whether it was due to population expansion, which affects the whole genomes of individuals.

Gene trees and the ethnogenesis process

Most genetic approaches to the reconstruction of population history are oversimplifications and in some cases there has been confusion between inferences about genes and inferences about populations, treating the latter as single evolutionary units. Quite often, a classification of populations has been interpreted as an evolutionary tree, ignoring that population dynamics follows a much more complex pattern than a simple split-and-diverge process. The examples of ethnogenesis shown by Moore (1994) are excellent illustrations of the real complexity that a population reconstruction can disclose. As complexity is likely to be high, reconstruction of history is bound to be an oversimplification of reality.

Recent studies using detailed information on DNA sequences or similar sets of data represent a quantum leap in the reconstruction of population history. In this case, like mtDNA, Y-chromosome and sequences for a given region in a large sample of individuals, it is possible to superimpose two events. At one level there is the molecular diversification of a genetic region, in which new mutations occur and the whole evolutionary structure may be reconstructed in what is called a gene tree. At another level there is population history, including spread, expansions and migrations that took place, while individuals, through generations, carried their genes in their process of diversification. Recent numerical tools allow reconstruction of the tree and estimation of the ages of the formation of new branches confined to specific geographic areas. Using such work it is possible to reconstruct the population history in a time frame and recognise migration not only at the population level, but also at the individual level (Comas et al. 1998; Bertranpetit 2000 and other recent papers).

The present development of genetics is providing new possibilities for understanding our evolutionary history and, in a more restricted time frame, the history of populations and their structure. In fact, among the six potential benefits listed in Human Genome Project Research one includes 'bioarchaeology, anthropology, evolution, and human migration' (http://www.ornl.gov/hgmis/project/about.html). In recent years we have seen results from the biological study of humans expanding to be of interest in other fields, and this will, no doubt, increase. Once we know all the possibilities of genetic and biological analysis it will be a good time to ask interesting questions about ourselves. After all, it takes good questions to stimulate the quest for answers.

References

Alves-Silva, J., Santos, M.S., Guimarães, P.E.M., Ferreira, A.C.S., Bandelt, H.-J., Pena, S.D.J. and Prado, V.F. (2000). The ancestry of Brazilian mtDNA lineages. *American Journal of Human Genetics*, **67**, 444–461.

Barbujani, G., Magagni, A., Minch, E. and Cavalli-Sforza, L.L. (1997). An apportionment of human DNA diversity. *Proceedings of the National Academy of Sciences, USA*, **94**, 4516–4519.

Bertranpetit, J. (2000). Genome, diversity and origins: the Y chromosome as a storyteller. *Proceedings of the National Academy of Sciences, USA*, **97**, 6927–6929.

Calafell, F., Comas, D., Pérez-Lezaun, A. and Bertranpetit, J. (2000). Genetics and population history of Central Asia. In *Archaeogenetics: DNA and the Population Prehistory of Europe*, ed. C. Renfrew and K. Boyle, pp. 259–266. McDonald Institute Monographs. Cambridge: Cambridge University Press.

Carvalho-Silva, D.R., Santos, F.R., Rocha, J. and Pena, S.D.J. (2001). The phylogeography of Brazilian Y-chromosome lineages. *American Journal of Human Genetics*, **68**, 281–286.

Cavalli-Sforza, L.L. and Bodmer, W.F. (1971). *Genetics of Human Populations*. San Francisco: W. Freeman.

Comas, D., Calafell, F., Mateu, E., Pérez-Lezaun, A., Bosch, E., Martínez-Arias, R., Clarimon, J., Facchini, F., Fiori, G., Luiselli, D., Pettener, D. and Bertranpetit, J. (1998). Trading genes along the silk road: mtDNA sequences and the origin of central Asian populations. *American Journal of Human Genetics*, **63**, 1824–1838.

Diamond, J.M. and Rotter, J.I. (1987). Observing the founder effect in human evolution. *Nature*, **329**, 105–106.

Dobzhansky, T. (1972). Natural selection in mankind. In *The Structure of Human Populations*, ed. G.A. Harrison and A.J. Boyce. Oxford: Oxford University Press.

Gagneux, P. and Varki, A. (2001). Genetic differences between humans and great apes. *Molecular Phylogenetics and Evolution*, **18**, 2–13.

Graur, D. and Li, W.H. (2000). *Fundamentals of Molecular Evolution*. Sunderland, MA: Sinauer Associates.

Human Gene Mutation Database: http://archive.uwcm.ac.uk/uwcm/mg/hgmd0.html

Human Genome Project Research: http://www.ornl.gov/hgmis/project/about.html

Kimura, M. (1983). *The Neutral Theory of Molecular Evolution*. Cambridge: Cambridge University Press.

Krings, M., Salem, A., Bauer, K., Geisert, H., Malek, A.K., Chaix, L., Simon, C., Welsby, D., Di Rienzo, A., Utermann, G., Sajantila, A., Pääbo, S. and Stoneking, M. (1999). mtDNA analysis of Nile river valley populations: a genetic corridor or a barrier to migration? *American Journal of Human Genetics*, **64**, 1166–1176.

Moore, J.H. (1994). Putting anthropology back together again: the ethnographic critique of cladistic theory. *American Anthropologist*, **96**, 925–948.

Pérez-Lezaun, A., Calafell, F., Seielstad, M., Mateu, E., Comas, D., Bosch, E. and Bertranpetit, J. (1997). Population genetics of Y-chromosome tandem repeats in humans. *Journal of Molecular Evolution*, **45**, 265–270.

Pérez-Lezaun, A., Calafell, F., Comas, D., Mateu, E., Bosch, E., Martínez-Arias, R., Clarimon, J., Fiori, G., Luiselli, D., Facchini, F., Pettener, D. and Bertranpetit, J. (1999). Sex-specific migration patterns in Central Asian populations, revealed by analysis of Y-chromosome short tandem repeats and mtDNA. *American Journal of Human Genetics*, **65**, 208–219.

Seielstad, M.T., Minch, E. and Cavalli-Sforza, L.L. (1998). Genetic evidence for a higher female migration rate in humans. *Nature Genetics*, **20**, 278–280.

Stoneking, M. (1998). Women on the move. *Nature Genetics*, **20**, 219–220.

6

Social institutions and demographic regimes in non-industrial societies: a comparative approach

RICHARD SMITH

In most non-industrial societies the key social institution determining the character of the demographic regime is the family or kinship system. In studying such an influence on the demographic characteristics of agrarian societies of the past or the present, historians of the family, social anthropologists and sociologists have frequently been badly served by the quality of the demographic data available. The collection of vital records is frequently a development that emerges once a society is well advanced on the path of economic development. Historical demography has therefore attracted the attention of students of demographic systems in non-industrial societies working in a variety of disciplines, since it has forged a methodology that has generated high quality data using the records of ecclesiastical parishes.

As a sub-discipline of the social sciences, historical demography has generated findings disproportionately from past European societies, since its methodology has been heavily dependent upon exploitation of a particular class of records relating to the registration of baptisms, marriages and burials by the parochial clergy of the Christian Church – a practice that was widely pursued from the sixteenth century (Mols 1954). Historical demography has also been able to assemble evidence from those parts of the world which were colonised by Europeans who carried with them registration practices that had long characterised their homelands. Investigations of past demographic conditions in areas that were not so completely colonised by Europeans are more restricted in scope, dependent on use of different types of evidence or on the registrations of events made by missionaries for those sections of societies that were subsequently Christianised. In this chapter the focus will be principally on demographic data that derive from the exploitation of parish registers in Europe from the period prior to the large-scale adoption of marital fertility control from the late nineteenth century. However, some reference will be made to evidence from non-European areas, particularly in contexts where marital fertility control was, or is, restricted and the economy was, or is, dominated by agriculture. These comparisons might be regarded as legitimate since the distinguished French demographer, Louis Henry (1956, 1967), who may be regarded as the founding father of historical demography, developed a set of procedures for the exploitation of seventeenth and eighteenth century French parish registers. This was initially to investigate situations dominated by what he termed *natural fertility* or the absence of parity-dependent fertility control. The particular technique we owe to Henry is known as *family reconstitution*: this uses nominal linkage of baptisms, marriages and burials to create demographic measurements relating to individuals within families. This information could not be secured until regular national censuses and vital registration systems were established, in most cases only from the late nineteenth century (Wrigley 1966).

The principal aim of the chapter will be to identify key features of the social structure that are reflected in the family systems which account for and give meaning to the demographic patterns that may be regarded as constituting demographic systems in Europe before the onset of the Demographic Transition. The ideas of Thomas Malthus have served as an intensively discussed demographic model for evaluating the links between a distinctive type of social structure and a particular constellation of demographic patterns in Europe *c.* 1550–1850. This chapter falls into two main parts. It begins with an assessment of the Malthusian model with particular reference to the familial and associated social contexts of those areas of Europe to which that model is thought to have been most applicable. It then proceeds, in a second shorter section, to consider demographic evidence from certain non-European areas with the intention of establishing certain demographic features that are thought to be strongly associated with joint family arrangements. It will also consider whether Malthus's ideas concerning those regions are borne out by evidence, that is now available, which Malthus did not have at his disposal. Throughout the chapter the aim will be to consider the extent to which demographic systems can be specified by reference to an amalgam of parameters relating to nuptiality, fertility, mortality and net migration, which in turn have distinctive geographical associations with family and kinship variables.

The Malthusian model

Malthus (1798) in his celebrated *Essay on the Principle of Population* identified two kinds of checks to population growth. Either population growth was contained by restricting nuptiality, which Malthus termed the *preventive check*; or, in the absence of this check, population would grow to the point where mortality rose and returned population to levels more appropriate for the resource basis of the society concerned (*positive check*: see Figure 6.1). Malthus was probably the first to argue that the accommodation reached between population and resources was influenced not only by the distribution of wealth in society and the way this determined individual economic well-being, but he was also aware of the significance of value systems which, as Schofield (1989: 279) has stressed, affected 'inter-personal relations within the family and the wider collectivity'. In the second edition of the *Essay* (1803) Malthus emphasised that the preferred route to the containment of population was via the preventive check, operating through 'moral restraint' as individuals showed an unwillingness to marry until they were in a position to support a family utilising their own resources. For Malthus, such preferences would stimulate savings, diminish poverty and thereby preserve incomes at levels substantially above subsistence, and hence act as a catalyst for general happiness and prosperity. Such behaviour, Malthus believed, effectively disabled the operation of the positive check. He wrote, in the second edition of the *Essay*, that

> in comparing the state of society which has been considered in this second book with that which formed the subject of the first, I think it appears that in modern Europe the positive checks to population prevail less, and the preventive checks more than in past times and in the more uncivilised parts of the world
>
> (1803: 315)

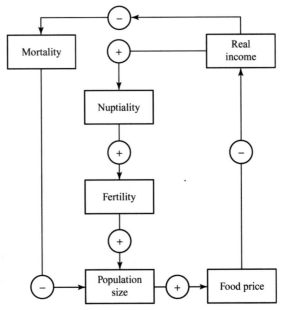

Figure 6.1. The positive and preventive checks.

Such views, when linked by modern historical demographic re-searchers to their findings on various aspects of European marriage in the period before 1800, have given rise to the idea that the growing af-fluence of Western societies from that date owed much to the ways in which they behaved demographically (Macfarlane 1978, 1986, 1987).

Malthus specifically identified China as a society whose demogra-phy was dominated by the positive check since it supposedly revealed little evidence of a preventive check. To China modern commenta-tors have added India as classic instances of 'uncivilised' parts of the world on the assumption that their marriage regimes were devoid of the moral restraint that was to be found in Europe. These character-isations of India and China have proved contentious (Goody 1996; Lee and Feng 1999).

Which features of European marriage in the past do historical demographers active in the last 30 years regard as particularly sup-portive of Malthus's ideas as they were formulated in the second edition of his *Essay* in 1803? In a seminal essay, Hajnal (1965) identi-fied what he believed was a distinctive European marital attribute. He

argued, on the basis of evidence from the seventeenth to the nineteenth centuries, that women in that region were distinctive in a wider
world setting because they married at a relatively late age (24–27
years on average) and a substantial number (between 5% and 15%)
never married at all. These suggestions were subsequently buttressed
by the mass of data assembled by the Princeton European Fertility
Project, which published a summary statement of its combined findings assembled over more than two decades from the 1960s (Coale
and Treadway 1986). That project provided the first fully inclusive
statistical cross-section of European nuptiality from data sources concentrated upon the year 1870 in the form of an index, *Im*. That index
is a measure of the contribution of marital status to the overall rate
of childbearing; strictly speaking, it is a measure from a hypothetical
population, in which only married women are fertile, and in which
married women are subject to maximum fertility rates at each age. *Im*
is the ratio of the number of births to married women in such a population to the number that would be produced if all women were married.
It measures the extent to which marital status would limit childbearing if marital fertility were natural and non-marital fertility were zero.
The geographical distribution of *Im* values provided vivid confirma-
tion of Hajnal's thesis, published over 20 years previously, that a fault
line running from Trieste to St Petersburg separated a 'European'
marriage pattern to the west from a non-European pattern to the
east. To the west of the line *Im* values were generally below 0.55 and
rose steadily to the east, exceeding 0.7 in much of Russia and the
Balkans – a feature these areas shared with the Middle East, North
Africa, and South and East Asia. Hajnal (1965) was firmly of the view
that a core zone for the European marriage pattern existed in north-
west Europe – a region containing England, northern France, the Low
Countries, most of Scandinavia (excluding Finland) and the western
regions of Germany. To the south of the Alps and the Pyrenees there
were suggestions that marriage for females was earlier and fewer
women avoided it altogether, as in 31 out of 48 Spanish provinces
Im values exceeded 0.8 and in 12 out of 16 Italian provinces *Im* was
over 0.55.

In a subsequent article Hajnal (1982) emphasised two other features that were characteristic of the core area of what he termed

north-west European household formation. These were that marriage was for the broad mass of the population neolocal in that the bride and groom set up their new residence apart from that of their respective natal families, frequently in a different community. A second and highly important feature of this marital residence pattern was that a newly married couple would be totally, or largely, economically independent of their kin. In the contrast that Hajnal drew between the household formation systems in north-west European areas and those in many other parts of the world, he emphasised the greater costs incurred by the newly marrying couples in the former area. In establishing their households, they were not absorbed within pre-existing households but had to carry all of the start-up costs associated with acquiring housing and stocking that household with material possessions. To obtain the resources needed to sustain an independent household, young adults might have to spend a lengthy time accumulating savings. Hajnal (1965) noted that in north-west Europe the interval between puberty and eventual marriage was passed by both males and females as servants or apprentices in the households of those to whom they were generally not related and from whom they received payment in the form of bed, board and an annual stipend. This last was frequently retained by the employer/household head until the service or apprenticeship contract was completed. Hajnal (1982) believed that this institution gave flexibility to marriage timing since, in difficult years or periods, the service 'net' expanded to incorporate many who had been unable to acquire the necessary resources to embark on marriage. He also noted that the resources that would be needed by a newly married couple to head their own household could also be secured through access to a farm or a craft workshop. Such resources might be obtained through either a post-mortem inheritance or a pre-mortem redistribution of assets within the family in the form of marriage gifts or settlements. Whichever means of resource acquisition applied, a bride and groom under Hajnal's north-west European marriage system would marry neolocally. Grooms therefore generally became heads of household at first marriage. In fact the assumption of household headship was concentrated within a comparatively narrow age range, between 25 and 30 years. For example, in England from the sixteenth to mid-nineteenth

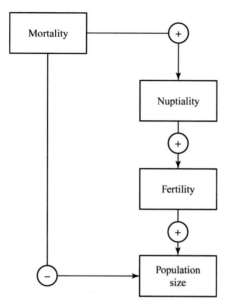

Figure 6.2. A system of ecological niches.

centuries the crude rate of first marriages per thousand persons aged 15 to 34 gives an approximate surrogate measure for the rate of new household creation per thousand persons aged 15–34 (Wrigley and Schofield 1989). The pattern exhibited by this measure is very similar to that displayed by the *Gross Reproduction Rate*, suggesting that the incidence of first marriage, the 'nuptiality valve' as Lesthaeghe (1980) termed this feature of European reproductive behaviour, was determining the secular shifts in fertility, before the onset of widespread contraception within marriage (see Figure 6.2).

Demographic homeostasis in West European peasant society

As Lesthaeghe (1980: 532) noted, 'the story of the nuptiality valve in western Europe before 1850 is well known', with a sizeable component of the female reproductive capacity under-exploited or unexploited because of the relatively late age of entry into, and a significant number of women remaining definitively out of, marriage. It has frequently

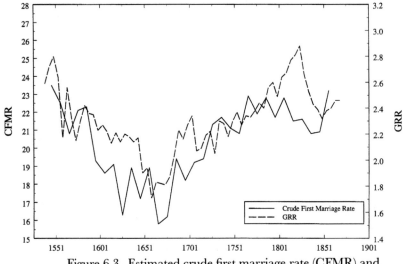

Figure 6.3. Estimated crude first marriage rate (CFMR) and gross reproduction rate (GRR) (marriages per 1,000 persons aged 15–34).

been asserted that this nuptiality pattern acted as a safety valve in the creation of demographic homeostasis responding as a dependent variable to shifts in mortality. Lesthaeghe argues that the 'central force in demographic homeostasis is the force of mortality' (1980: 528). Such a conceptualisation of homeostasis treats fertility as marching to the tune played by mortality (see Figure 6.3). If mortality is assumed to have been unstable, showing no detectable secular trend, *ipso facto* nuptiality is the principal 'driver' of fertility. If mortality improved then fertility would decline accordingly to sustain a stationary demographic outcome (Dupâquier 1972). Such a conceptualisation of the relationship between mortality and fertility has been deeply influential in the work of continental European historical demographers. A well-known case study that exemplifies such a viewpoint is that by Dupâquier (1979). His research on the Paris basin in northern France in the period *c.* 1650–1800 emphasises a demographic equilibrium that was continually re-establishing itself despite disturbances largely initiated by epidemics or exogenously determined mortality movements. The nuptiality valve opened and closed in response to the mortality rises and falls such that marriage age and marital in-cidence adjusted to compensate for the losses or savings of lives that occurred. For much of the late seventeenth and the eighteenth cen-tury, Dupâquier (1979) argues, the number of hearths barely changed at all in the Paris basin. To explain this phenomenon, use is made of the concept of an agricultural holding or craft workshop as fulfilling a function analogous to that of a territory in a bird population in which a new breeding pair is only allowed to establish itself once a nest is vacated. Such a system assumes a strong controlling role fulfilled by older generations over their younger subordinates. This purgatory was often relieved in the short term as epidemics struck and provided those in junior positions with an opportunity to establish a new house-hold (Dupâquier 1972). There is little doubt that in the eighteenth century French society as a whole, a largely peasant agrarian eco-nomy, displayed the attributes of the homeostatic regime *par excellence* with fertility declining as life expectancy increased especially after 1740 (Wrigley & Schofield 1989).

A similar relationship between mortality and fertility may have ap-plied more widely over northern parts of Spain and north-central Italy where evidence suggests that notable shifts in marital patterns took place in the eighteenth and nineteenth centuries (Livi-Bacci 1978; Perez Moreda 1986; Rowland 1988). In all of these latter regions case studies suggest a pattern of early marriage (18 to 22 for female first marriages) and of low proportions never marrying in the years be-fore and after 1350, which is still identifiable into the early eighteenth century. However, *Im* values, as revealed by the Princeton Fertility Project (Coale and Treadway 1986), suggest that by the nineteenth century there had been a significant diminution in the incidence of female marriage. In all these regions mortality fell and the low nup-tiality that prevailed at the end of the eighteenth century was, as Livi-Bacci (2000: 106) notes, 'the culmination of a process initiated in the sixteenth century'. Livi-Bacci (2000) is convinced that in all of the regions to the east of the 'Hajnal line' a high-nuptiality system prevailed in the late fourteenth and fifteenth centuries in reaction to devastating plague losses and a breaking down of the economic ob-stacles that restricted access to marriage. Such a view is, of course, in accordance with Lesthaeghe's (1980) promotion of mortality change as a driving force in the shift from high to low nuptiality, and from a high- to low-pressure demographic regime.

The nuptiality valve in the English case before the mid-nineteenth century

This concept of a nuptiality valve took its cue from shifts in mortality. While it is of central significance as a demographic accounting framework for large parts of the pre-industrial world, it is of very limited value in understanding the English case from the sixteenth to the mid-nineteenth centuries (Smith 1981, 1990). In England, shifts in nuptiality and in associated fertility (Gross Reproduction Rate) appear in the main not to have operated primarily as a short-term equilibrating force which reacted immediately and principally in the wake of mortality surges; rather, shifts in fertility and mortality took the form of long, often century-long, waves of growth and decline (see Figure 6.4). The mortality series, as represented by expectation of life at birth, also suggests a cycle which, while less regular than that displayed by fertility, reveals a lengthy deterioration through the seventeenth century. This was followed by a recovery to late sixteenth century levels by the early nineteenth century. Long-run changes in mortality were not systematically related to changes in real wages; but long-run changes in a crude first marriage rate were. Rather than homeostasis, the pattern suggests that fertility behaviour may have been responding systematically to secular changes in society and economy, that were in no sense captured by the changing access to a fixed number of slots, in a manner so elegantly described by Dupâquier (1972, 1979) for France as a whole or the inhabitants of the Paris basin in particular.

What may have been the sources of disequilibrium in the English demographic system? One such source may have been the institution that formed a central plank in Hajnal's north-west European household formation system. In England, through most of the sixteenth, seventeenth and eighteenth centuries, a majority or a very large minority of each new cohort of young men and women in their teens and early twenties hired themselves out for annual contracts,

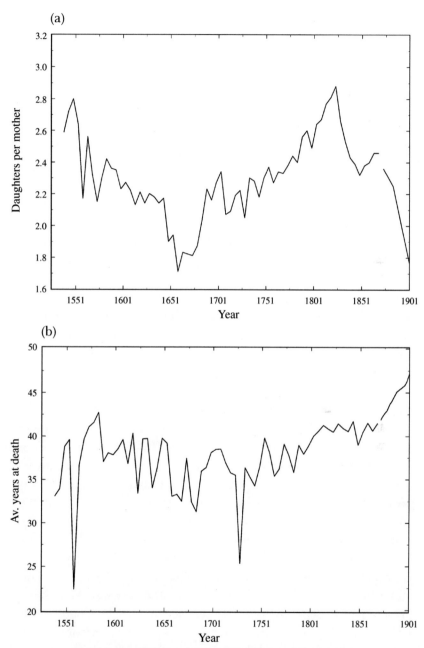

Figure 6.4. (a) Changes in English fertility, 1536–1901: quinquennial GRRs. (b) Changes in English mortality, 1536–1901: quinquennial e_0s.

whereby they laboured in the fields and workshops and acquired skills that they might use in later life. While departure from the natal hearth was commoner among children of the relatively poor, it was not absent among the wealthy echelons of society. Consequently, it was not necessarily economics that caused parents to put their children out to service, but acceptance of a social norm, i.e. a view of what constituted the normal transitional stage between dependent childhood at home and independent existence later in life (Kussmaul 1981). It should be stressed that this inter-household movement of children brought about a divergence between the private and the social rate of return on children. It led to some parents paying the costs of rearing their own children, but also receiving the economic return from other people's children (Smith 1986). Furthermore, the existence of the institution of life-cycle service created 'dysfunctional' effects on demographic trends. If attention is placed upon the demand rather than the supply factors that determined the incidence of service, the reasons for this dysfunction can be readily appreciated. In England it would seem that demand for servant labour on the part of farmers increased when population growth was zero or negative and wages were relatively high due to labour shortages. Under such conditions farmers preferred to employ live-in servant labour paid disproportionately in board and lodging. However, when wages were low and population growing, farmers preferred to hire their labour by the day and remunerate with cash rather than kind. Consequently servants in England tended to increase in their incidence when mortality was high and to exacerbate the demographic impact of reduced life expectancies by reducing nuptiality and hence fertility in the younger adult age groups (Smith 1981).

The English early modern economy, like that of Holland, was far more differentiated and urbanised than that of the remainder of Europe before the nineteenth century (de Vries and van der Woude 1997; Smith 2000). The inheritance of land was not the principal means by which individuals secured access to a livelihood (Smith 1984a). Children frequently left the parental hearth around the age of 14 or 15 years and, when later they married, in three-quarters of all instances the new couple resided in a place in which neither partner had been born (Schofield 1971). A large share of overall employment was in work for wages. The family was not a key provider of support during episodes of illness, unemployment or in old age (Smith 1984b). Much more important as a source of assistance was the community in the form of parish-funded welfare. The family was therefore not the mediating force through which inheritance to a farm or workshop provided a conduit in which mortality and fertility were inversely related. Furthermore, England was an open society with a mobile population, into which new diseases were introduced and circulated through an urban system with London at its core. Emigration from the rest of England and into London disturbed the geographical distribution of the sexes and facilitated the emergence of long periods when women were drawn into the labour force as substitutes for men, thereby not marrying when economic circumstances might have been favourable for them to do so (Smith 1992).

Some demographic correlates of joint household formation systems

Although contrasts have been drawn between the presence and absence of fertility-determined demographic homeostasis in two different variants of the west European system of household formation associated with female late marriage, these pale in significance when comparisons are drawn between these systems and others associated with joint household formation systems and early female marriage. The most important distinguishing feature of the joint household systems is that they are not associated with marital practices that give rise to a newly married couple establishing a separate household (Davis 1955; Hajnal 1982). Rather, the newly married become elements in pre-existing households. It would, however, be unwise to treat joint household formation systems as undifferentiated. Such systems are extensively distributed through east and south Asia. For instance, the North Indian joint family system produces a situation where sons tend to remain with the parents. Daughters marry out and the parents continue to exercise control over household affairs until death. Daughters marry out at young ages but, especially among the landowning castes, the joint family does not extend the opportunity to marry to all sons;

this is in order to protect against the buildup of pressure on household resources (Das Gupta 1997). In such a patrilineal system the position of those in-marrying females is potentially hazardous (Das Gupta 1995a). In the North Indian joint household system there are strong inter- and intra-generational bonds between household members related to each other by blood. Women marrying into such households are in a weak position to protect their own and their children's health, resulting in maternal and infant mortality rates which are higher than those found in systems in which married couples manage their own affairs. Restricted husband–wife communication and authority vested heavily in the senior household members have significant implications for fertility behaviour. There is a strong preference for sons as they are so critical to securing the future well-being of adult females in the household. Under such conditions mothers develop strong bonds with their sons, and daughters can be the subject of discrimination by their mothers and by other household members (Cain 1986). As a result mortality rates among girls are significantly higher than among boys. This is a feature lacking in north-west European household systems in which bilateral kinship prevailed, and the conjugal bond was strong relative to any other kinship links both within and between households (Das Gupta 1995b).

In India, far higher proportions of elderly persons live with their children than would have been found in north-west European areas. When co-residence occurred in the latter region, it was often precarious with the elderly couple or parent moved to a separate building or room on the edge of the farmhouse; a marginal or subordinate relationship to the child who had taken over household headship was then the norm. In contrast, Indian parents living with their offspring typically exercise far greater autonomy and would expect to play a central, perhaps dominant, role in household affairs. It might be expected that in north-west Europe the elderly would have experienced relatively poorer health and have displayed higher mortality relative to adults of prime working age. However, substantial redistribution of resources from communal funds to deal with their demotion within the family may have eradicated such a possibility. In fact, the absence of communally managed welfare funds is more likely to be found in areas of the North Indian household systems. Women without sons would be especially disadvantaged, showing far higher mortality than those able to co-reside with sons (Cain 1986). Furthermore, infant mortality rates were likely to have been lower in the north-west European households since more family resources were likely to move to the young as a result of the lower claims of the elderly within the family than would be the case in joint households (Das Gupta 1997).

The traditional Chinese household formation system and its demographic correlates

The pro-natalist implications of the familial relationships in North Indian households have been frequently stressed. However, it is important not to assume that joint household systems inevitably led to higher overall fertility than those of western and northern Europe. In China, where another variant of the joint household system prevailed, universal female marriage was encountered until very recently (Lee and Feng 1999). Yet, fertility may not have been higher than in Europe since fertility in marriage seems to have been distinctly low. Western married women in the period before *c.* 1850, had they experienced the marital fertility rates of a given year, would have produced seven to nine births. Married women in China would have only produced six or fewer children (Zhao 1997a, b). How far this was a reflection of infanticidal practices which depressed the number of children recorded as having been born to an individual mother is currently a matter of some debate among historical demographers. There is no doubt that the Chinese did practise noteworthy female infanticide (Chen 1989). Recent historical investigations suggest that in the eighteenth century approximately 10% of female births resulted in infanticide and, when this is combined with neglect of baby girls, created female infant mortality rates that were frequently between two and three times higher than those found in north-west Europe before 1800 (Lee 1981; Lee *et al.* 1994). Furthermore, because of the depressed fertility and enhanced infant mortality, Chinese parents resorted relatively frequently to fictive kinship and adoption to overcome the limitations of biology and miscalculation, and preserve family continuity and support in old age (Wolf and Huang 1980).

Recently students of traditional Chinese demographic systems have contrasted those systems with European practices which they would claim left the question of nuptiality to solely human agency (Lee and Feng 1999). The Chinese household formation systems gave rise to demographic behaviour which resulted in a wider array of choices that included the decision to give away or kill children and the adoption of others' children. However, Malthus had originally argued that such practices made it possible for Chinese females to avoid marital restraint. Malthus wrote:

> This permission thus given to parents thus to expose their offspring tends undoubtedly to facilitate marriage, and encourage population. Contemplating this extreme resource beforehand, less fear is entertained of entering into the married state
>
> (1803: 129)

Malthus probably would not have denied that Chinese patterns of demographic behaviour rested heavily on collectivist decision-making within the household. In fact some commentators on the present demographic setting in China, which arises largely from the remarkably successful adoption of a one-child policy, stress that this transition is not the result of any fundamental ideological transformation. It is just the continuation of collectivist goals, in which the control over decision-making is shifted from the family, especially the older generation, to the state. The result has made it possible for China to experience the swiftest fertility transition in recorded history (Lee and Feng 1999).

West Africa

In West Africa matrifocal households, embedded within lineage formation, have created another highly distinctive demographic regime. In this region marriage is a gradual process and can be easily terminated, thereby increasing the probability of a woman having several partners over her lifetime (Meekers 1992). Each marital arrangement creates expectations that the woman will bear children, which makes it difficult for her to gain any overall sense of how many children she might have. Furthermore, there is no assumption that the mother of a child will carry the responsibility for its upbringing. This care is just as likely to be provided by other members of the lineage through fostering arrangements. These reduce the likelihood that couples will carry the full costs of their reproductive acts, leading to relatively high rates of total fertility (Goody 1973; Isiugo-Abanihe 1985; Bledsoe 1990). None the less, this region is notable for long periods of post-partum taboo on sexual relations. However, it has been argued that such behaviour becomes understandable within the framework of lineage structures in which it is the desire of elders to prevent the formation of close conjugal relationships (Caldwell and Caldwell 1981). Lesthaeghe has argued that

> control of the reproductive pool and of female and child labour are only two among many elements of an overall pattern of gerontocratic control over people in lineages ... The prevention of close husband–wife solidarity through the long post-partum taboo on sexual intercourse in particular, and through the maintenance of physical and psychological distance in general, also ensures lineage dominance.
>
> (1980: 530)

Conclusion

This chapter began by considering certain features of those European societies that by the late eighteenth century Malthus supposed to have acquired distinctive demographic characteristics because of their unique marriage practices which ensured that fertility was far from maximised. The chapter has tried to show that family and kinship systems have a variety of means of influencing both fertility and mortality, and that the north-west European system, about which Malthus wrote so approvingly, was not the only one to limit population growth without succumbing to the painful impact of the positive check (see also Attenborough, Chapter 10). The Malthusian system was founded upon the creation of neolocal marriages which gave rise to only one offspring group per family. It was by no means a universal system and had little relevance to areas in which joint familes prevailed. Furthermore, family systems affect not only fertility, but also infant and child mortality, the circulation of children and young adults, the demographic prospects of those who survive to maturity, the gender balance of whole populations and the composition of migrant streams. 'Family system norms, then impinge directly on every term within the demographic equation' (Skinner 1997: 85).

References

Bledsoe, C. (1990). The politics of children: fosterage and the social management of fertility among the Mende of Sierra Leone. In *Births and Power: Social Change and the Politics of Reproduction*, ed. W.P. Handwerker, pp. 81–100. Boulder, CO: Westview Press.

Cain, M. (1986). The consequences of reproductive failure: dependence, mobility and mortality among the elderly of rural South Asia. *Population Studies*, **40**, 375–388.

Caldwell, P. and Caldwell, J. (1981). The function of child-spacing in traditional societies and the direction of change. In *Childspacing in Tropical Africa: Traditions and Change*, ed. H. Page and R. Lesthaeghe, pp. 73–92. New York: Academic Press.

Chen, G. (1989). Songdai shengai buyu fengsu de shengxing jiqi yanyi [The reasons for the use of infanticide during the Song]. *Zhongguo shi yanjiu [Research in Chinese History]*, **1**, 138–143.

Coale, A.J. and Treadway, R. (1986). A summary of the distribution of overall fertility, marital fertility and the proportion married in the provinces of Europe. In *The Decline of Fertility in Europe*, ed. A.J. Coale and S.C. Watkins, pp. 31–181. Princeton, NJ: Princeton University Press.

Das Gupta, M. (1995a). Life course perspectives on women's autonomy and health outcomes. *American Anthropologist*, **97**, 481–491.

Das Gupta, M. (1995b). Fertility decline in the Punjab: parallels with historical Europe. *Population Studies*, **49**, 481–500.

Das Gupta, M. (1997). Kinship systems and demographic regimes. In *Anthropological Demography: Toward a New Synthesis*, ed. D. Kertzer and T. Fricke, pp. 36–52. Chicago: University of Chicago Press.

Davis, K. (1955). Institutional patterns favouring high fertility in underdeveloped areas. *Eugenics Quarterly*, **2**, 33–39.

De Vries, J. and van der Woude, A. (1997). *The First Modern Economy: Success, Failure and Perseverance of the Dutch Economy, 1500–1815*. Cambridge: Cambridge University Press.

Dupâquier, J. (1972). De l'animal à l'homme: le mécanisme autorégulateur des populations traditionelles. *Révue de l'Institut de Sociologie*, **11**, 177–209.

Dupâquier, J. (1979). *La Population Rurale du Bassin Parisien à l'époque de Louis XIV*. Paris: Presses Universitaires de France.

Goody, E. (1973). *Contexts of Kinship: An Essay on the Family Sociology of the Gonja of Northern Ghana*. Cambridge: Cambridge University Press.

Goody, J. (1996). *The East in the West*. Cambridge: Cambridge University Press.

Hajnal, J. (1965). European marriage patterns in perspective. In *Population in History: Essays in Historical Demography*, ed. D.V. Glass and D.E.C. Eversley, pp. 101–147. London: Edward Arnold.

Hajnal, J. (1982). Two kinds of pre-industrial household formation system. *Population and Development Review*, **8**, 449–484.

Henry, L. (1956). *Anciennes Familles Genevoises: Etude Démographique, XVIe–XXe Siècle*. Paris: Presses Universitaires de France.

Henry, L. (1967). *Manuel de Démographique Historique*. Paris: Libraire Droz.

Isiugo-Abanihe, U. (1985). Child fosterage in West Africa. *Population and Development Review*, **11**, 53–73.

Kussmaul, A. (1981). *Servants in Husbandry in Early Modern England*. Cambridge: Cambridge University Press.

Lee, B. (1981). Infanticide in China. In *Women in China: Current Directions in Historical Scholarship*, ed. R. Guisso and S. Johannesen, pp. 163–177. Lewiston, NJ: Edwin Mellen Press.

Lee, J. and Feng, W. (1999). Malthusian models and Chinese realities: the Chinese demographic system 1700–2000. *Population and Development Review*, **25**, 33–66.

Lee, J., Feng, W. and Campbell, C. (1994). Infant and child mortality among the late imperial Chinese nobility: implications for two kinds of positive check. *Population Studies*, **48**, 395–411.

Lesthaeghe, R. (1980). On the social control of human reproduction. *Population and Development Review*, **6**, 527–548.

Livi-Bacci, M. (1978). *La Société Italienne Devant la Mortalité*. Florence: Dipartimento Statistico.

Livi-Bacci, M. (2000). *The Population of Europe*. Oxford: Basil Blackwell.

Macfarlane, A. (1978). *The Origins of English Individualism: Family, Property, and Social Transformation*. Oxford: Basil Blackwell.

Macfarlane, A. (1986). *Marriage and Love in England: Modes of Reproduction 1300–1840*. Oxford: Basil Blackwell.

Macfarlane, A. (1987). *The Culture of Capitalism*. Oxford: Basil Blackwell.

Macfarlane, A. (1997). *The Savage Wars of Peace: England, Japan and the Malthusian Trap*. Oxford: Basil Blackwell.

Malthus, T.R. (1798). *An Essay on the Principle of Population* [Volume 1 of *The Works of Thomas Robert Malthus*, ed. E.A. Wrigley and D. Souden (1986)]. London: William Pickering.

Malthus, T.R. (1803). *An Essay on the Principle of Population* [Volume 2 of *The Works of Thomas Robert Malthus*, ed. E.A. Wrigley and D. Souden (1986)]. London: William Pickering.

Meekers, D. (1992). The process of marriage in African societies: A multiple indicator approach. *Population and Development Review*, **18**, 61–78.

Mols, R. (1954). *Introduction à la Démographie Historique des Villes d'Europe du XIVe au XVIIIe Siècle*, 3 Vols. Louvain: Université de Louvain.

Perez Moreda, V. (1986). Matrimonio y familia algunas considerraciones sobre el modelo matrimonial español en la edad moderna. *Boletin de la Asociacion de Demografia Historica*, **4**, 3–51.

Rowland, R. (1988). Sistemas matrimoniales en la Peninsula Iberica (siglos xvi–xix). Una perspectiva regional. In *Demografica Historica en España*, ed. V. Perez-Moreda and D.S. Reher, pp. 74–137. Madrid: Ediciones el arquero.

Schofield, R.S. (1971). Age-specific mobility in an eighteenth-century rural English parish. *Annales de Démographies Historiques*, **5**, 261–274.

Schofield, R.S. (1989). Family structure, demographic behaviour and economic growth. In *Famine, Disease and the Social Order in Early Modern Society*, ed. J. Walter and R. Schofield, pp. 279–304. Cambridge: Cambridge University Press.

Skinner, G.W. (1997). Family systems and demographic processes. In *Anthropological Demography: Towards a New Synthesis*, ed. D.I. Kertzer and T. Fricke, pp. 53–95. Chicago: University of Chicago Press.

Smith, R.M. (1981). Fertility, economy and household formation in England over three centuries. *Population and Development Review*, **7**, 595–622.

Smith, R.M. (1984a). Some issues concerning families and their property in rural England. In *Land, Kinship and Life-cycle*, ed. R.M. Smith, pp. 1–86. Cambridge: Cambridge University Press.

Smith, R.M. (1984b). The structured dependence of the elderly as a recent development: Some sceptical historical thoughts. *Ageing and Society*, **4**, 409–428.

Smith, R.M. (1986). Transfer incomes, risk and security: the roles of the family and the collectivity in recent theories of fertility change. In *The State of Population Theory*, ed. D.A. Coleman and R. Schofield, pp. 188–211. Oxford: Basil Blackwell.

Smith, R.M. (1990). Monogamy, landed property and demographic regimes in pre-industrial Europe: regional contrasts and temporal stabilities. In *Fertility and Resources*, ed. J. Landers and V. Reynolds, pp. 164–188. Cambridge: Cambridge University Press.

Smith, R.M. (1992). Influences exogènes et endogènes sur le 'frein préventif' en Angleterre, 1600–1750: quelques problèmes de spécification. In *Modèles de la Démographies Historique*, ed. A. Blum, N. Bonneuil and D. Blanchet, pp. 175–192. Paris: Presses Universitaires de France.

Smith, R.M. (2000). Relative prices, forms of agrarian labour and female marriage patterns in England, 1350–1800. In *Marriage and Rural Economy: Western Europe since 1400*, ed. I. Devos and L. Kennedy, pp. 19–48. Turnhout: Brepols.

Wolf, A. and Huang, C. (1980). *Marriage and Adoption in China 1845–1945*. Stanford, CA: Stanford University Press.

Wrigley, E.A. (1966). Family reconstitution. In *An Introduction to English Historical Demography*, ed. D.E.C. Eversley, P. Laslett and E.A. Wrigley, pp. 96–159. London: Weidenfeld and Nicolson.

Wrigley, E.A. and Schofield, R.S. (1989). *The Population History of England 1541–1871: A Reconstruction*. Cambridge: Cambridge University Press.

Zhao, Z. (1997a). Demographic systems in historic China: Some new findings from recent research. *Journal of the Australian Population Association*, **14**, 201–232.

Zhao, Z. (1997b). Deliberate birth control under a high-fertility regime: reproductive behaviour in China before 1970. *Population and Development Review*, **23**, 729–767.

7

The dynamics of child survival

EMILY K. ROUSHAM AND LOUISE T. HUMPHREY

Introduction

Changes in child survival rates are very significant in the dynamics of human populations, and mortality is an important variable in studies of demography, genetics and anthropology. Furthermore, rates of child survival provide an important indication of the general biological welfare of a population.

Children are a particularly vulnerable group, both physically and socially, within human populations (Caldwell 1996). This vulnerability is reflected in mortality rates, which are highest during infancy, decrease from 1 to 4 years and reach a minimum from 5 to 9 years of age. The physical vulnerability of children stems from an immature immune system, rendering them particularly susceptible to infectious diseases. This is coupled with a high energy and nutrient requirement relative to body weight. These intakes are required to fuel the high energy demands of the human brain, as well as for somatic growth and development. One of the most critical periods for child survival is the weaning period, typically from six months onwards, when infants make the transition from dependence on mother's milk to reliance on the local diet (Martorell 1995). Children are also socially vulnerable, being entirely dependent on adult carers for the first year of life and remaining highly dependent up to the age of five. This social dependence means that children are susceptible to conscious or unconscious neglect (Caldwell 1996).

This chapter examines the sociocultural and biological correlates of child survival, contrasting populations with low and high survival rates. Pre-industrialised societies typically have high rates of infant and child mortality. The majority of deaths are due to infectious diseases, often in combination with undernutrition. Industrialised populations experience much lower rates of infant and child mortality disease and far fewer deaths from infectious disease. Instead, mortality is more commonly due to chronic, non-infectious diseases or lifestyle diseases that typically strike in late adulthood. The change in disease patterns observed as populations go through the demographic transition is commonly referred to as the epidemiological transition. These closely linked processes continue to have a marked impact on child survival, life expectancy and the demographic structure of populations.

Populations with high child mortality

This section describes the patterns of survival in populations with high child mortality. In particular, it outlines the effects of disease, malnutrition and sociocultural influences on child survival. High child mortality is widespread in the Less Developed Countries (LDCs) of the world. Many characteristics of survival in LDCs today would also have been seen in pre-industrialised European populations of the past.

The leading cause of death in LDCs is infectious disease, accounting for an estimated 14 million deaths of children under the age of five each year (UNICEF 1996). This can be further broken down by specific disease categories: diarrhoea is estimated to cause 35% of deaths of children under five years and malaria 21% of deaths. Measles and acute respiratory infections are the next most common causes of death. In reality, however, children do not die of a single infectious agent, and it is often the combination of illnesses, or the exacerbation of their condition because of the underlying presence of malnutrition, which leads to death.

The effect of infectious disease is far worse in the presence of childhood malnutrition. The synergism between disease and nutritional status is responsible for many deaths in the developing world (Pelletier *et al.* 1993). Measles provides a classic example of this interaction. In Western populations, measles is a mild disease of childhood and rarely presents with complications. In developing countries, measles is a severe, and often fatal, infection. This contrasting severity of the same viral infection is largely due to the nutritional status of the host (Morley 1980; Tomkins and Watson 1989). In Africa, where wasting and underweight are highly prevalent, measles typically affects children at a younger age. The rash is more severe, affecting all the epithelial surfaces of the body including the respiratory and gastrointestinal tracts. This leads to complications such as laryngitis, bronchopneumonia and diarrhoea (Morley 1980; Noah 1988). Inadequate treatment for these complications quickly leads to dehydration and weight loss, and possible death. Nutritional improvements alone could therefore prevent many deaths from measles.

Some argue that greater emphasis on improvements in nutritional status will lower infant mortality (Schroeder and Martorell 1997) and there is growing evidence that nutritional interventions can reduce morbidity and mortality from disease. For example, vitamin A supplementation can lead to lower child mortality (Herrera et al. 1992). Likewise, improved zinc status reduces morbidity from diarrhoeal and respiratory infections (Tomkins 2000). Breastfeeding provides immune protection and this in turn lowers the incidence of diarrhoea and respiratory infection in infants. In sum, many childhood deaths are attributed to acute infectious diseases when, in fact, malnutrition is the underlying or predisposing factor (Tomkins 2000). To redress this balance, strategies must aim to reduce the prevalence of undernutrition as well as lowering infectious disease morbidity.

With high rates of child mortality, the genetic structure of a population would be shaped by deaths from infectious disease. One of the few well-documented examples of particular genotypes conferring resistance to a disease is that of malaria, caused by the organism Plasmodium falciparum. Genetic resistance to malaria infection has been selected for in the form of sickle-cell haemoglobin (HbS), several other haemoglobinopathies, glucose-6-phosphate dehydrogenase (G-6-P-D) deficiency and the Duffy antigen (Marsh 1993).

The risks of death due to disease and malnutrition are, of course, compounded by poverty and socioeconomic status. Disadvantaged groups in living and in past populations have poorer access to health facilities, lower educational attainment, and often require their children to work from an early age. These groups have extremely low food security and are particularly vulnerable during times of famine or seasonal food shortages.

Social and cultural values also have a distinct effect on child survival. In some cases, cultural factors reinforce socioeconomic inequalities. In India, social inequalities are maintained over successive generations by the caste system and this is reflected in child mortality differentials. Ethnicity can also influence rates of child survival. A recent analysis of survey data from 11 sub-Saharan African countries found significant differences in mortality of children under five years between ethnic groups. The differences were closely linked with economic inequality between ethnic groups in many of the countries studied (Brockerhoff and Hewett 2000).

Excess mortality may be linked to very specific cultural practices. For example, one study in rural Nepal found that children delivered in an animal shed were at significantly higher risk of dying in the first year of life than those born at home. The difference was caused by inadequate preparation of the animal sheds prior to delivery, which facilitated infection of both the newborn and the mother (Thapa *et al.* 2000). Examples such as this emphasise the need for strategies that are designed to reduce child mortality within the framework of prevailing cultural norms.

Culturally determined attitudes can also result in selective discrimination or conscious neglect of particular groups of children. In many countries of the world and particularly in parts of the Middle East, South Asia and the Far East, excess mortality of female children occurs (Waldron 1987). This represents a reversal of the excess male mortality seen in Western countries, and is results from sociocultural phenomena rather than any biological predisposition. The causes of excess female mortality stem ultimately from a preference for sons, often in combination with patriarchal social structures and a low status of women. The proximate determinants of excess female mortality are not always apparent, but there is evidence that girls receive poorer dietary intake and health care (Chen *et al.* 1981; Koenig and D'Souza 1986).

Poor health of female children may not be universal within populations with a preference for boys. A longitudinal study of growth and nutrition among Bangladeshi children found that gender bias was apparent only among the poorer, landless households. Among wealthier, farming households girls and boys were equal in growth and nutritional status (Rousham 1999). It was also observed that undernutrition in landless girls was greatest during food shortages, showing a temporal dynamic to the risk of mortality (Rousham 1999).

An extreme example of gender discrimination has been seen as a result of the one-child policy in China. In the early 1980s, the male:female ratio at birth was 116:100 in rural China, compared with the expected ratio of 106:100. This very masculine sex ratio undoubtedly came about through female infanticide, unreported female births and sex-selective abortions. In 1995, the sex ratio at birth for rural areas was reported to be 108:100, suggesting that the relaxation of population control measures has reduced bias against girls (Hesketh and Zhu 1997). After birth, however, there is still evidence of son preference and excess female mortality (Ren Xinhua 1995). This drastic population control programme will no doubt have long-term consequences on the gender balance in China.

In other circumstances too, sociocultural values can affect the survival of children according to their age. In rural Peru, children under one year of age are valued less highly than older children and therefore warrant less attention and care (Larme 1997). Similarly, sons are valued more highly than daughters and are afforded special treatment, particularly when ill. This ethnographic study therefore sheds light upon the underlying causes of selective neglect against infants and females. In the absence of birth control measures, this neglect may be a means of regulating family size and sex ratio (Larme 1997).

The decline of infant and child mortality

The process of industrialisation in Europe and elsewhere has been accompanied by a dramatic reduction in infant and child mortality, much of which is due to the decline of infectious disease mortality. However, the specific factors responsible for this decline are not easy to identify. Following the Industrial Revolution there were radical improvements in public health brought about by the provision of clean water, improved housing and social welfare. Given that so many aspects of employment, agriculture, housing and patterns of residence also changed simultaneously, it is difficult to attribute the decline of infectious disease to any single factor. In Britain, medical intervention made a late contribution to the control of infectious diseases: immunisations began long after the decline and antibiotic therapy has only become readily available over the last 50 years. McKeown (1976) highlighted the relatively small contribution of medical intervention to the decline of infectious disease. This is illustrated by the graphs showing death rates of children under 15 years from whooping cough and measles over the last 150 years (Figure 7.1). Clearly, much of the decline took place long before either vaccination or the use of antimicrobials.

Information on the causes of death and factors affecting mortality in populations undergoing industrialisation is very limited. The burials from the crypt of Christ Church, Spitalfields in London, however, provide a very rare insight into mortality profiles over the period 1729–1852 (Molleson and Cox 1993). The Spitalfields sample represents a transitional population, in that it shares features of both pre- and post-industrialised populations. The families in the sample were relatively wealthy craftsmen and professionals, but many experienced crowded living and working conditions, and large family sizes and child labour were not uncommon. Mortality rates under the age of 15 years were 20.5% in the Spitalfields sample compared with 45–57% for London as a whole (Molleson and Cox 1993). Child survival was therefore considerably higher in this sample than among poorer populations of London at the time. The sample also shows evidence of greater male mortality than female. Figure 7.2 shows that 26% of males died before the age of 15 compared with 16.8% of females. More specifically, male mortality doubled that of females from 3–12 months and from 2–10 years. The first interval corresponds to the time of weaning and the second period may be due to differences in the work, home or recreational activities of boys and girls. Interestingly, previous studies based on nineteenth-century parish records have also reported excess male mortality in childhood (Woods and Hinde 1987).

In sum, at the time of industrialisation and a decline in child mortality, the Spitalfields sample provides evidence of social class differences

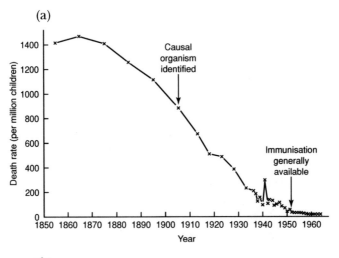

Figure 7.1. Death rates of children under 15 in England and Wales from (a) whooping cough (b) measles. Source: McKeown (1976).

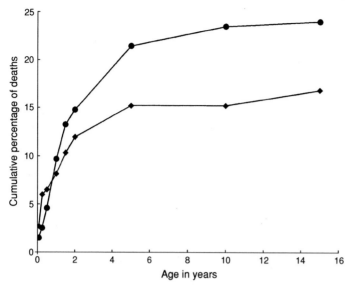

Figure 7.2. Cumulative percentage of deaths between birth and 15 years for males (•) and females (♦) from the Spitalfields coffin plate sample.

in mortality, and of excess male mortality during childhood. These patterns are still apparent in contemporary Britain and are discussed in the next section.

The process of industrialisation is associated not only with a decline in child mortality, but also with a marked reduction in fertility. Three of the theories that have been proposed as explanations for this change are the changing economic value of children, increased autonomy of women and changing expectations of infant mortality. A recent analysis of data from developed societies between 1880 and 1940 and developing countries between 1960 and 1990 found that infant mortality was the best predictor of fertility, and that female empowerment was also a good predictor (Sanderson and Dubrow 2000). Nevertheless, some of the underlying determinants of fertility have so far proved to be resistant to demographic and epidemiological change. The sex of surviving children is still an important determinant of subsequent fertility in countries throughout the world. In Korea, for example, women with a son are less likely to have another child and take longer to conceive any subsequent child (Larsen *et al.* 1998), and in Botswana, male preference for sons still influences both desired family size and eventual fertility (Campbell and Campbell 1997). In contrast, analysis of fertility patterns in Denmark has demonstrated a strong preference for a balanced sex ratio, and higher fertility rates in two-boy families than in two-girl families indicate a moderate preference for girls (Jacobson *et al.* 1999).

Strategies to reduce child mortality in developing countries have changed in the light of various successes and failures. In the 1960s, the World Health Organization (WHO) held optimistic plans to eradicate many infectious diseases, heralded by the global campaign to eradicate smallpox, which was successfully achieved in 1977 (Fenner 1980). Campaigns to reduce other diseases such as hookworm infection, yellow fever and malaria were, however, unsuccessful. The difficulty of eradicating disease has prompted the WHO to shift its priorities towards broader primary health care measures such as immunisation and the provision of oral rehydration therapy (ORT). These two programmes alone are estimated to prevent *c.* 3.5 million deaths each year (UNICEF 1996). Global immunisation against measles, polio, typhoid, whooping cough and diphtheria reached an average of 80% of children in 1990 (UNICEF 1996). Despite the success of immunisation, there is still no vaccine for diarrhoea or for malaria, which continue to cause millions of deaths each year (Tomkins 2000).

Paradoxically, economic development and westernisation do not necessarily eliminate selective discrimination against female children. Despite declining fertility rates in India, sex ratios at birth are increasingly biased towards males (Basu 1999). Analysis of demographic data suggests that selective abortion, infanticide or unreported births are taking place among certain sectors in India (Basu 1999; Sudha and Rajan 1999). A smaller average family size provides fewer opportunities to have sons and selection against daughters (either before or after birth) may be just as great, despite the overall decline in child mortality in India (Sudha and Rajan 1999).

In terms of population structures, the demographic and epidemiological transitions signify a reduction in selective pressures at early ages. Infectious disease mortality, however, does not necessarily end with economic development. The appearance of HIV/AIDS and the resurgence of tuberculosis in both developed and developing countries demonstrate the continuing threat of disease to child survival. In the case of AIDS, the demographic and social effects in some parts of the world are already dramatic.

Populations with low child mortality

During the past century, there have been great improvements in infant and child survival in England and Wales. Infant mortality is now around six per thousand births, raising the question of whether any further reduction is possible (Schuman 1998).

In the context of low overall mortality, there are marked differences in mortality depending on gender, social class, or genetic and behavioural factors. In most Western nations, males have higher mortality throughout infancy, childhood and adolescence. For example, boys in England and Wales are almost twice as likely as girls to die from sudden infant death syndrome (SIDS) in the first year of life (ONS 1998). This may be due to immaturity of the male respiratory system or greater susceptibility of male infants to respiratory infections. Excess male mortality may also stem from the possession of only one X chromosome and a lower immune resistance to infectious diseases. In later childhood, males die disproportionately from accidents and injuries. Among children aged 1–14 years in England and Wales, *c.* 33% of all deaths among boys are due to injury and poisoning compared with only 20% of deaths among girls (ONS 1998). This clearly reflects social and behavioural differences between the sexes, with young men more likely to practise risk-taking behaviour. The cumulative effect of excess male mortality is a shorter life expectancy for men compared with women. Interestingly, the greatest reduction in infant mortality over the last ten years is due to a decrease in SIDS, which accounted for 49% of post-neonatal deaths in 1988 and 26% of deaths in 1996. This has come about through a simple educational campaign to modify parental behaviour rather than any medical breakthrough.

A strong association between infant mortality and social class is still evident in recent data from England and Wales. In 1995, infant

mortality rates were 68% higher in social class V (unskilled occupations) than in social class I (professional occupations) (Schuman 1998).
Infant mortality was also 25–35% higher among babies born outside
marriage compared with those born inside marriage. Ethnic differences in infant mortality cannot be ascertained from birth and death
registrations, but children of mothers who were born outside the UK
had 23% higher infant mortality than those whose mothers were
born within the UK. The highest mortality rates were among births
to mothers who were born in Pakistan or in the Caribbean (Schuman
1998). These examples demonstrate the persistence of social inequalities in a Western industrialised nation. These social inequalities are
translated into differential mortality rates, which have a disproportionate effect on lower socioeconomic classes.

Advances in medical science and technology have also influenced
child survival in industrialised populations. Just two examples of this
are the use of assisted reproductive technologies to overcome low
fertility and neonatal intensive care to aid the survival of premature
babies. In fact, the numbers of births conceived through *in vitro* fertilisation (IVF) and of premature births are too small to affect the overall
population structure, but they have led to significant increases in the
survival of multiple births as well as altering common norms and
expectations of reproduction and child survival.

The availability of prenatal testing and screening techniques has
also reduced the number of children born with genetic or congenital
abnormalities. In many cases, such as severe neural tube defects, this
merely leads to the earlier detection of non-viable foetuses, which
would otherwise have ended in stillbirths or neonatal deaths. Therapeutic abortion for conditions such as Down's Syndrome may lead to
a significant reduction of cases in the population. On the other hand,
those who are born with the condition may receive far better medical care and support within the community than previously. Some
genetic conditions no longer present a threat to survival because of
successful diagnosis and treatment. For example, all newborns are
screened for phenylketonuria (PKU) and those diagnosed with the
condition receive dietary intervention to limit the intake of the amino
acid phenylalanine. This is sufficient to prevent the development
of severe mental retardation that would otherwise occur soon after

birth. Gene therapy has yet to become a realistic option for many genetic anomalies, but therapeutic treatments for conditions such as haemophilia and cystic fibrosis are being vigorously pursued. In sum, the effects of medical technologies on specific gene frequencies vary according to the nature of the condition, the success of diagnosis and the availability of treatment. In many cases these changes are too small to have any overall effect on the gene frequencies of the population as a whole.

Early survival and later morbidity

Events that occur in the earliest stages of an individual's life are important not only for their effect on development and survivorship during infancy and childhood but also for their longer term implications for morbidity and mortality. A large body of research has demonstrated compelling links between health in early life and later adulthood (Barker 1994). Much of this has focused on relatively affluent Westernised populations. In these groups, the foetal environment has been shown to influence the incidence of non-transmissible diseases such as heart disease and non-insulin dependent diabetes mellitus in later life (Barker *et al.* 1990; Phipps *et al.* 1993; Leon *et al.* 1998). Other research suggests that the effects of an impoverished early environment can also influence mortality rates during early adulthood (Moore *et al.* 1997; Stodder 1997). A study of survivorship in three villages in Gambia, where seasonal variation in exposure to inadequate diet and disease is marked, demonstrated that people born during the hungry season were more susceptible to fatal infections during early adulthood than those born in the harvest season (Moore *et al.* 1997). Stodder (1997) analysed the frequency of enamel hypoplasias in a human skeletal sample from the Mariana Islands and found that those who died over the age of 21 years had fewer enamel defects than those who failed to survive into adulthood. Since enamel hypoplasias are caused by an interruption of the normal process of enamel formation, these results indicate that individuals who survived into adulthood experienced fewer episodes of developmental disruption during the period of tooth crown formation.

Techniques for examining the relationship between events in early life and subsequent morbidity and mortality are very different in living and past populations (Humphrey and King 2000). Studies of modern populations rely on hospital or medical records, which give details of the circumstances surrounding birth, and observations of current medical status (e.g. Leon *et al.* 1998). For past populations, it is usually necessary to infer both age at death and experience of stress in early life from dental and osteological evidence. The Spitalfields sample is an unusual example of a past population since parameters that would normally be inferred osteologically are known from historical records (Molleson and Cox 1993) and accurate information on age at death and date of birth is available for 50 individuals who died aged over 15 years.

Seasonality

A further component to child mortality in temperate regions is explored in the following discussion, where the relationship between season of birth and age at death is examined. The Spitalfields adults were divided into two groups, reflecting those born during the colder winter months (November–April) and those born during the warmer summer months (May–October). The mean age at death for adults born in summer was 58 years compared with a mean age at death of 48 years for those born in winter. Although seemingly large, this difference is not significant ($p = 0.114$) owing to small sample sizes and the large amount of overlap between the two groups. Seasonal variation in diet and disease is less extreme in Britain than in many other parts of the world. Nevertheless, winter may have been a time of greater hardship in eighteenth- and nineteenth-century London, with reduced availability of fresh foods and a higher incidence of respiratory diseases.

Even though the majority of LDCs are not in temperate zones of the world, there are seasonal differences in climate that affect health and nutrition in many ways and, although the conditions are not directly comparable, the information from the Spitalfields material provides a clearer indication of factors to consider than is usually available.

Summary

This chapter has highlighted some of the causes of death in infancy and childhood among both living and past populations, as well as describing some of the features associated with survival and how these have changed. Childhood survival depends upon a combination of the genetic makeup of the individual and the environment into which that individual is born. Poverty, social disadvantage and disease are strong selective agents in both contemporary and historical populations. These may shape the genetic and demographic structure of the population from one generation to another. Through such mechanisms populations 'adapt', so that the interaction of genes and the environment in subsequent generations plays an important role in programming individuals to the conditions *in utero* and in early infancy. In turn, this interaction in early childhood shapes the morbidity and mortality of children and adults, thereby continuing the cycle of effects on the population and its dynamics.

References

Barker, D.J.P. (1994). *Mothers, Babies and Disease in Later Life.* London: British Medical Journal Publishing Group.

Barker, D.J.P., Bull, A.R., Osmond, C. and Simmonds, S.J. (1990). Fetal and placental size and risk of hypertension in adult life. *British Medical Journal*, **301**, 259–262.

Basu, A.M. (1999). Fertility decline and increasing gender imbalance in India, including a possible south Indian turnaround. *Development and Change*, **30**, 237–263.

Brockerhoff, M. and Hewett, P. (2000). Inequality of child mortality among ethnic groups in sub-Saharan Africa. *Bulletin of the World Health Organisation*, **78**, 30–41.

Caldwell, P. (1996). Child survival: physical vulnerability and resilience in adversity in the European past and the contemporary third world. *Social Science and Medicine*, **43**, 609–619.

Campbell, E.K. and Campbell, P.G. (1997). Family size and sex preferences and eventual fertility in Botswana. *Journal of Biosocial Science*, **29**, 191–204.

Chen, L.C., Huq, E. and D'Souza, S. (1981). Sex bias in the allocation of food and health care in rural Bangladesh. *Population and Development Review*, **7**, 55–70.

Fenner, F. (1980). Smallpox and its eradication. In *Changing Disease Patterns and Human Behaviour*, ed. N.F. Stanley and R.A. Joske, pp. 215–229. London: Academic Press.

Herrera, M.G., Nestel, P., Elamin, A., Fawzi, W.W., Mohamed, K.A. and Weld, L. (1992). Vitamin-A supplementation and child survival. *Lancet*, **340** (8814), 267–271.

Hesketh, T. and Zhu, W.X. (1997). The one child family policy: the good, the bad, and the ugly. *British Medical Journal*, **314**, 1685–1687.

Humphrey, L.T. and King, T. (2000). Childhood stress: a lifetime legacy. *Anthropologie*, **38**, 33–49.

Jacobson, R., Moller, H. and Engholm, G. (1999). Fertility rates in Denmark in relation to the sexes of preceding children in the family. *Human Reproduction*, **14**, 1127–1130.

Koenig, M.A. and D'Souza, S. (1986). Sex differences in childhood mortality in rural Bangladesh. *Social Science and Medicine*, **22**, 15–22.

Larme, A. (1997). Health care allocation and selective neglect in rural Peru. *Social Science and Medicine*, **44**, 1711–1723.

Larsen, U., Chung, W. and Das Gupta, M. (1998). Fertility and son preference in Korea. *Population Studies – A Journal of Demography*, **52**, 317–325.

Leon, D.A., Lithell, H.O., Vågerö, D., Koupilová, I., Mohsen, R., Berglund, L., Lithell, U.-B. and McKeigue, P.M. (1998). Reduced fetal growth rate and increased risk of death from ischaemic heart disease: cohort study of 15000 Swedish men and women born 1915–29. *British Medical Journal*, **317**, 241–245.

Marsh, K. (1993). Immunology of human malaria. In *Bruce-Chwatt's Essential Malariology*, ed. H.M. Gilles and D.A. Warrell, pp. 60–77. London: Edward Arnold.

Martorell, R. (1995). Promoting healthy growth: rationale and benefits. In *Child Growth and Nutrition in Developing Countries: Priorities for Action*, ed. P. Pinstrup-Anderson, D. Pelletier and H. Alterman, pp. 15–31. New York: Cornell University Press.

McKeown, T. (1976). *The Modern Rise of Population*. London: Edward Arnold.

Molleson, T. and Cox, M. (1993). *The Spitalfields Project*, Volume 2. *The Middling Sort*. Council for British Archaeology Research Report 86.

Moore, S.E., Cole, T.J., Poskitt, M.E., Sonko, B.J., Whitehead, R.G., McGregor, I.A. and Prentice, A.M. (1997). Season of birth predicts mortality in rural Gambia. *Nature*, **388**, 434.

Morley, D. (1980). Severe measles. In *Changing Disease Patterns and Human Behaviour*, ed. N.F. Stanley and R.A. Joske, pp. 115–127. London: Academic Press.

Noah, N. (1988). Measles. In *Elimination or Reduction of Disease*, ed. A.J. Silman and S.P.A. Allwright, pp. 46–59. Oxford: Oxford University Press.

Office of National Statistics (1998). *Mortality Statistics: Childhood, Infant and Perinatal, England and Wales, 1996*. Series DH3 No. 29. London: The Stationery Office.

Pelletier, D.L., Frongillo, E.A. and Habicht, J.P. (1993). Epidemiologic evidence for a potentiating effect of malnutrition on child mortality. *American Journal of Public Health*, **83**, 1130–1133.

Phipps, K., Barker, D.J.P., Hales, C.N., Fall, C.H., Osmond, C. and Clark, P.M. (1993). Fetal growth and impaired glucose tolerance in men and women. *Diabetologia*, **36**, 225–228.

Ren Xinhua, S. (1995). Sex differences in infant and child mortality in three provinces in China. *Social Science and Medicine*, **40**, 1259–1269.

Rousham, E.K. (1999). Gender bias in South Asia: effects on child growth and nutritional status. In *Sex, Gender and Health*, ed. T.M. Pollard and S.B. Hyatt, pp. 37–52. Cambridge: Cambridge University Press.

Sanderson, S.K. and Dubrow, J. (2000). Fertility decline in the modern world and in the original demographic transition: testing three theories with cross-national data. *Population and Environment*, **21**, 511–537.

Schroeder, D.G. and Martorell, R. (1997). Enhancing child survival by preventing malnutrition. *American Journal of Clinical Nutrition*, **65**, 1080–1081.

Schuman, J. (1998). Childhood, infant and perinatal mortality, 1996; social and biological factors in deaths of children aged under 3. *Population Trends*, **92**, 5–14.

Stodder, A.L.W. (1997). Subadult stress, morbidity and longevity in Latte Period populations on Guam, Mariana Islands. *American Journal of Physical Anthropology*, **104**, 363–380.

Sudha, S. and Rajan, S.I. (1999). Female demographic disadvantage in India 1981–1991: sex selective abortions and female infanticide. *Development and Change*, **30**, 585–618.

Thapa, N., Chongsuvivatwong, V., Geater, A.F., Ulstein, M. and Bechtel, G.A. (2000). Infant death rates and animal-shed delivery in remote rural areas of Nepal. *Social Science and Medicine*, **51**, 1447–1456.

Tomkins, A. (2000). Malnutrition, morbidity and mortality in children and their mothers. *Proceedings of the Nutrition Society*, **59**, 135–146.

Tomkins, A.M. and Watson, F.E. (1989). *Malnutrition and Infection: A Review*. Geneva: World Health Organisation.

United Nations Children's Fund (1996). *The State of the World's Children*. New York: Oxford University Press.

Waldron, I. (1987). Patterns and causes of excess female mortality among children in developing countries. *World Health Statistics Quarterly*, **40**, 194–210.

Woods, R. and Hinde, A. (1987). Mortality in Victorian England: models and patterns. *Journal of Interdisciplinary History*, **18**, 27–54.

8

Genetic structure of south Indian caste populations: a confluence of biology and culture

L.B. JORDE, M.J. BAMSHAD, W.S. WATKINS,
C.E. RICKER, M.E. DIXON, B.B. RAO,
B.V.R. PRASAD AND J.M. NAIDU

The Indian subcontinent, which contains nearly one-sixth of the world's population, has received migrants from many sources during its history. Its population is consequently diverse, both culturally and genetically (Nei and Roychoudhury 1982; Majumder and Mukherjee 1993; Majumder 1998). The patterns of this diversity can offer important clues to evolutionary history. In addition, unique sociocultural phenomena in India have influenced mating patterns for thousands of years. It is expected that these patterns have in turn affected the genetic structure of the population. For these reasons, the Indian population is an appealing subject for population genetic analysis. This chapter, while concentrating on some populations in southern India, shows how modern genetic techniques can play an important part in our understanding of human population dynamics.

Three major waves of immigration have influenced the genetic structure of India. Palaeolithic immigrants are thought to have originated from Africa, or possibly Australia, and may have contributed to the gene pool of present-day tribal populations (Maloney 1974; Chandler 1988; Majumder and Mukherjee 1993; Cavalli-Sforza *et al.* 1994). A second major wave of proto-Dravidian-speaking immigrants came from the Fertile Crescent area *c*. 10,000 years ago and populated most of the subcontinent (Cavalli-Sforza *et al.* 1994). The third major wave, emanating from west-central Asia about 3,500 years ago, consisted of Indo-European-speaking 'Caucasoid' populations. This immigration event is marked archaeologically by the appearance of

Painted Grey Ware (Thapar 1980). The Indo-European speakers also introduced iron, the domestic horse, the Aryan languages, a mixed pastoral and agrarian economy and, it seems, the patrilineal kinship system. The relatively recent entrance of this population into northern India, with subsequent gradual migration southward, has resulted in significant genetic differences between the northern and southern populations of the subcontinent (Passarino *et al.* 1996a; Majumder 1998).

The Indo-European immigrants may have introduced the well-known caste system into India, although it is also possible that the caste system existed prior to their arrival (Karve 1968). More than 2,000 castes have been identified. Each caste was considered to be strictly endogamous, while lineages within a caste group were exogamous. The manifest for the caste system is found in the hymns of the *Rig Veda* in which the extant population was partitioned into four broader classes or *varnas*: the *Brahmins* (priests), the *Kshatriyas* (warriors), the *Vysyas* (traders), and the *Sudras*. It is thought that Aryan speakers from central Asia admixed with the upper varnas (i.e. Brahmins, Kshatriyas, and Vysya), while the Sudra descend from individuals with African, Southeast Asian and/or Australian admixture (Majumdar 1958; Balakrishnan 1978). A fifth varna, the *Pancham*, was added to include the ex-'untouchable' castes. Each varna, which is composed of a number of castes, or *jati*, was identified by occupation and associated with a defined social status. This social stratification was rationalised and strengthened by ritual and religious philosophy. It is commonly thought that the Indo-European speakers, having established or appropriated the caste system, tended to occupy its highest strata (Majumdar 1958; Balakrishnan 1978).

There is more regional and socioeconomic variability in the system than can be discussed in this chapter. In general terms, however, although castes are ideally endogamous, it has long been acknowledged that a small number of marriages and other matings between men and women from different castes do occur (Karve 1968). Typically, these unions occur between a man from a higher caste and a woman from a slightly lower caste, frequently through female hypergamy (Tambia 1973). Relevant to this chapter is the question of the status accorded to the offspring. The offspring of these hypergamous unions may join the father's caste, although their status may be somewhat less than that of their father. The net result is that female genetic lineages are more likely to move between castes than male lineages.

The majority of India's population (c. 82%) consists of Hindus. The caste system, though officially abolished in the 1960s, has strongly influenced the mating practices of this portion of the population. A second major component of the social structure of India's population consists of 'tribal' elements. Although the definition of 'tribe' appears to be somewhat arbitrary, it typically refers to populations considered to be aboriginal, inhabiting the Indian peninsula before the immigration of pastoral nomads from western and central Asia. These aboriginal populations are thought to descend from populations that expanded into South Asia during the Palaeolithic era. The size of different tribal groups varies from a few hundred (e.g. the Andamanese) to a few million individuals (e.g. the Gonds). Four hundred contemporary tribes constitute 7.5% of the total Indian population (Majumder and Mukherjee 1993; Sirajuddin *et al.* 1994).

India has been the focus of numerous biological anthropological studies using morphometrics (Majumder *et al.* 1990; Sirajuddin *et al.* 1994), dermatoglyphics (Singh 1978; Malhotra *et al.* 1980), protein polymorphisms (Roychoudhury 1974; Balakrishnan 1978, 1982; Char and Rao 1986; Char *et al.* 1989; Papiha *et al.* 1997), and mitochondrial DNA (mtDNA), Y chromosome and nuclear DNA polymorphisms (Mountain *et al.* 1995; Bamshad *et al.* 1996, 1998; Barnabas *et al.* 1996; Passarino *et al.* 1996a, b; Bhattacharyya *et al.* 1999; Kivisild *et al.* 1999; Majumder *et al.* 1999; Quintana-Murci *et al.* 1999a; Thangaraj *et al.* 1999; Watkins *et al.* 1999; Clark *et al.* 2000; Mastana *et al.* 2000). Although molecular genetic studies are now being pursued in India, most of the genetic studies published to date are based on blood groups and protein polymorphisms. From these studies, a few broad conclusions have emerged.

(i) Genetic variation is larger within castes and tribes than between them (Roychoudhury 1982; Das *et al.* 1996).
(ii) Substantial genetic differences are observed between caste and tribal groups (Das *et al.* 1996).

(iii) Genetic variation among tribal groups is extensive and is more structured geographically than linguistically (Murty *et al.* 1993; Sirajuddin *et al.* 1994; Papiha *et al.* 1997; Pitchappan *et al.* 1997).

(iv) Tribal populations are in general more similar to one another genetically than to the caste populations, but they are in turn more similar to caste populations than to major continental populations such as Africans or Australians (Roychoudhury 1984).

Although these studies have provided many useful insights, a more accurate and complete documentation of India's genetic structure awaits detailed, comparative studies using mitochondrial, Y chromosome and autosomal polymorphisms. The significance of studying mtDNA is that it is passed on maternally, and of studying the Y chromosome is that it is passed on paternally to males only. The autosomal material is inherited equally from each parent. Such analyses will also permit comparisons with data sets generated for other populations, leading to inferences about the origins of India's populations. With this objective in mind, we have analysed variation in mitochondrial, Y chromosome, and autosomal DNA polymorphisms in caste populations of South India.

Methods

Study populations

The populations analysed to date consist of 316 members of caste populations. Following groupings used in earlier studies of Indian castes (Krishnan and Reddy 1994), the populations can be divided according to status into upper (59 Brahmin, 10 Vysya, 11 Kshatriya), middle (58 Kapu, 53 Yadava, 28 Jalari, 23 Walabahija), and lower (19 Relli, 29 Madiga, 26 Mala). These individuals were sampled from the area around Visakhaptnam in the state of Andhra Pradesh. Fifteen tribal populations have also been sampled from South India (the tribal populations are not analysed in the present study). The importance of choosing populations from the south is that the ideals of isogamy, even cousin or classificatory cousin marriage, are stronger than in some parts of the north of India. All individuals sampled from the caste and tribal populations are unrelated males. Sampling was done with the approval of the Government of India and the Institutional Review Boards of the University of Utah and Andhra University.

For comparison, data were also included from 143 Africans (San, Sotho/Tswana, Mbuti Pygmy, Biaka Pygmy, Tsonga, Nguni, Hema, Alur and Nande populations), 120 Europeans (French, Poles, Finns, British and Scandinavians), and 61 members of Asian populations outside India (Malay, Vietnamese, Cambodian, Chinese and Japanese). The latter are hereafter termed simply 'Asians'. These populations, and the genetic data compiled for them, are described at greater length in previous publications (Jorde *et al.* 1995, 1997, 2000: Watkins *et al.* 2001).

Laboratory and statistical methods

Whole blood was collected by venous blood draw or finger prick. In most cases a hair sample was also collected. Samples were transported to Andhra University where DNA was extracted. The following DNA samples were used to generate the data reported in this study: 411 base pairs of mtDNA control region sequences, presence or absence of a 9-base pair deletion in the coding region of the mtDNA genome, six tandem repeat polymorphisms and three single nucleotide polymorphisms in the non-recombining portion of the Y chromosome, and 40 autosomal *Alu* polymorphisms. The latter polymorphisms are a type of 'mobile element' insertion; these have been inserted at random locations throughout the primate genome during the past 60 million years or so. Some of the most recently inserted *Alu* polymorphisms are specific to humans, and specific insertions are present in some individuals but absent in others. (For more detailed information on laboratory methods, see Appendix 8.1.)

Gene diversity, a concept to be used extensively in this chapter, refers to the amount of genetic variation in a population. If all members of the population differ from one another at a specific locus, then a typical gene diversity index would equal one for that locus. If all members are identical at the locus, the gene diversity index would equal zero. Genetic distances, also discussed at some length below, are a

Table 8.1. *Genetic distances between major caste groups*

Comparison	MtDNA	Y chromosome	*Alu*
Upper–middle	0.0037	0.0062	0.0069
Upper–lower	0.0149	0.0054	0.0181
Middle–lower	0.0067	0.0005	0.0071

measure of the extent of genetic differences between two populations. Typically, a genetic distance of 1 would indicate that two populations are maximally differentiated, while a distance of zero would indicate that the two populations are genetically identical (i.e. for the locus or group of loci in question). Gene diversity was estimated for each population for each type of genetic system (i.e. Y chromosome polymorphisms, mtDNA and autosomal *Alu* polymorphisms). Gene diversity levels within and between populations were used to estimate the proportion of genetic variance due to subdivision. Genetic distances between individuals and between groups were calculated. In addition, the correlation between inter-individual caste rank differences and inter-individual genetic distances was assessed. (For more detailed information on statistical methods see Appendix 8.1.)

Results

The matrices of inter-individual *Alu* genetic distances and differences in caste rank show a low but highly significant correlation using the Mantel test ($r = 0.07$, $p < 0.001$). This demonstrates significant genetic differentiation among individuals of different caste ranks, with greater differences generally occurring between individuals whose caste rank is more different.

The genetic distances between each pair of caste groups, as introduced above, are given in Table 8.1. For the mtDNA and *Alu* data, the genetic distance is larger between upper and lower caste groups than between either of these groups and the middle castes: social rank and genetic distance are correlated with one another. In contrast, caste

Table 8.2. *MtDNA genetic distances between caste groups and continental populations*

Caste group	Africa	Asia	Europe
Upper	0.1852	0.0452	0.0910
Middle	0.1844	0.0259	0.0785
Lower	0.1632	0.0232	0.1075

rank and genetic distance are not correlated for the Y chromosome polymorphisms.

The mtDNA genetic distances between each caste population and the major continental populations are given in Table 8.2 and depicted graphically in Figure 8.1. As in all mtDNA studies, Africans are the most divergent population. All three caste groups are genetically most similar to Asians and most dissimilar from Africans. A comparison of the distances between castes versus Africans and castes versus Europeans reveals an intriguing pattern: as one moves from upper to lower castes, the distance from African populations becomes progressively smaller, while the distance from European populations becomes progressively larger. This is consistent with relatively larger African genetic contributions to the lower castes and relatively larger European genetic contributions to the upper castes.

The Y chromosome distances, shown in Table 8.3, reveal a similar pattern. Again, the African distance becomes smaller as one moves from upper to lower castes, and the European distance is greater in the middle and lower castes than in the upper castes. In contrast to the mtDNA distances, the Y chromosome data do not demonstrate closer Asian affinity for all caste groups. The upper castes are in fact more similar to the Europeans, while the middle castes are slightly more similar to Europeans than to Asians. Only the lower castes show greatest affinity for Asian populations. As seen in Figure 8.2, the caste populations diverge from other world populations, indicating a possible male founder effect.

Similar to the Y and the mtDNA data, the *Alu* polymorphisms yield genetic distances that are smaller between the upper castes and European populations than between the lower castes and European

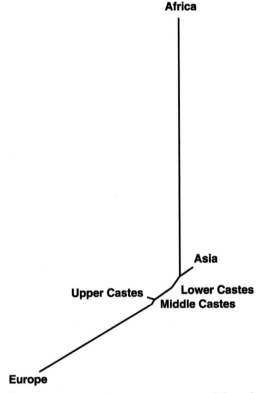

Figure 8.1. A neighbour-joining network based on mtDNA HVS1 sequence, portraying the genetic relationships among caste and continental populations.

populations (Table 8.4). In contrast to the mtDNA results but similar to those of the Y chromosome data, the affinity between upper castes and Europeans is higher than that between the upper castes and Asians. This pattern is also observable in Figure 8.3, which shows

Table 8.3. *Y chromosome genetic distances between caste groups and continental populations*

Caste group	Africa	Asia	Europe
Upper	0.0166	0.0104	0.0092
Middle	0.0156	0.0110	0.0108
Lower	0.0131	0.0088	0.0108

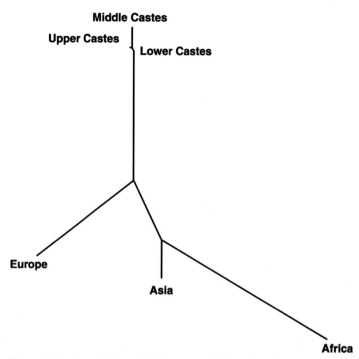

Figure 8.2. A neighbour-joining network based on Y chromosome STRs, portraying the genetic relationships among caste and continental populations.

that the upper castes lie on a branch of the tree leading to Europeans, while the middle and lower castes are on the branch leading to Asians.

Because 40 *Alu* polymorphisms were assessed, it is possible to assess the statistical significance of these results (because there is no recombination for the Y and the mtDNA polymorphism, each of these

Table 8.4. Alu *genetic distances (with standard errors) between caste groups and continental populations*

Caste group	Africa	Asia	Europe
Upper	0.074 ± 0.018	0.024 ± 0.009	0.011 ± 0.003
Middle	0.082 ± 0.018	0.013 ± 0.005	0.020 ± 0.006
Lower	0.083 ± 0.017	0.017 ± 0.005	0.027 ± 0.006

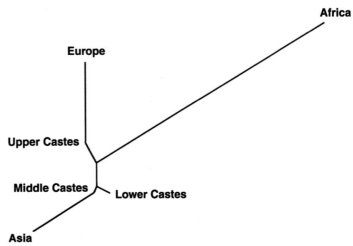

Figure 8.3. A neighbour-joining network based on autosomal *Alu* polymorphisms, portraying the genetic relationships among caste and continental populations.

sets of polymorphisms represents effectively a single locus, so statistical testing of genetic distances is futile). Historical evidence that the caste system was instituted by the relatively recent wave of migrants from West Eurasia would suggest greater affinity between the upper castes and Europeans than between the lower castes and Europeans. This involves two assumptions:

(i) The West Eurasian immigrants would be genetically more similar to present-day Europeans than to other members of the comparative populations, such as Asians. This assumption is reasonable because regional continuity appears often to have existed for human populations (Cavalli-Sforza *et al.* 1994).

(ii) The West Eurasian immigrants, having probably established the caste system, would tend to occupy its higher strata.

Under these assumptions, it is appropriate to use a one-tailed test of significance to examine the difference between the two genetic distances. The 90% confidence limits of Nei's standard distances estimated between upper castes and Europeans (0.006–0.016) versus lower castes and Europeans (0.017–0.037) do not overlap, indicating statistical significance at the 0.05 level. This result offers statistical support for a significantly greater affinity between upper caste and European populations than between lower caste and European populations.

Discussion

India's populations are noteworthy for their cultural, morphological and genetic (Nei and Roychoudhury 1982; Majumder and Mukherjee 1993) diversity. Using traditional blood group loci and protein polymorphisms, many previous studies have attempted to assess this diversity and to measure the affinities of India's populations with other world populations. These studies have provided evidence for affinities with a variety of other populations, including Africans, Asians, Australian Aborigines and Europeans (Roychoudhury 1977; Cavalli-Sforza *et al.* 1994). However, these results have been widely debated, and the relative degrees of similarity with these populations have remained controversial.

Some of these relationships are becoming clearer with the application of more powerful and sensitive molecular polymorphisms. mtDNA studies have consistently indicated that South Indian populations tend to be genetically most similar to Asian populations (Mountain *et al.* 1995; Bamshad *et al.* 1996; Passarino *et al.* 1996a), and comparisons of northern and southern Indian populations tend to indicate greater Asian affinity for the latter group (Passarino *et al.* 1996a). Our mtDNA results, which agree with these observations, are consistent with historical and archaeological data that indicate an early settlement of India by populations originating in Central Asia. Furthermore, the north–south cline in Asian influence is consistent with the more recent influx of Caucasoid peoples into North India, with a gradual percolation of their genes southward (Cavalli-Sforza *et al.* 1994).

The data presented here permit a more fine-grained analysis of the genetic relationships between major continental populations and specific caste groups. For mtDNA, all caste groups are most similar to Asians and least similar to Africans, but upper castes are relatively more similar to Europeans than are middle and lower castes. This may reflect the imposition of the caste system (and appropriation of the highest positions) by incoming Indo-European speakers 3,000–4,000 years ago (Majumdar 1958; Balakrishnan 1978). The lower caste group shows relatively greater similarity to Africans than do the other two groups, a result that is consistent with an earlier study based on a small sample of mtDNA HVS2 sequences (Bamshad *et al.* 1996). This result is again consistent with a settlement model in which the older populations of South India, with greater Asian and African genetic affinities, formed large portions of the lower castes while the upper castes tended to be dominated by Indo-Europeans.

Y chromosome variation in India remains poorly understood. The results presented here show that, as with mtDNA data, all caste groups are most distant from African populations. This is bolstered further by our finding that the Y-specific *Alu* polymorphism (DYS287, or YAP), which is relatively common in most African populations, was absent in all caste members. A similar result has been obtained in another recent study of Y chromosome variation in India (Thangaraj *et al.* 1999). Our Y chromosome results indicate that the male lineages in upper and middle caste groups are closest to Europeans, while those in the lower caste groups are most similar to Asians. This finding is supported somewhat by a recent study of Y chromosome variation that showed smallest genetic distances between Hindu and European populations (Quintana-Murci *et al.* 1999b). Furthermore, the *Alu* polymorphisms reveal a similar pattern, and they demonstrate significantly greater European affinity with upper castes than lower castes.

These findings offer an interesting contrast to the mtDNA results and may imply that the majority of immigrating Indo-European speakers were males who then mixed with females from the resident population. A similar suggestion has been put forward based on the fact that traditional blood group systems, which reflect both paternal and maternal lineages, often group Indian populations with European populations instead of Asians (Passarino *et al.* 1996b). Our result, which demonstrates the expected European affinities for male-specific lineages and also defines the pattern by caste group, strengthens this conjecture considerably. The hypothesis requires further testing with additional Y chromosome polymorphisms and autosomal DNA polymorphisms, and it needs to be tested and replicated in other Indian caste populations. It should be emphasised that these genetic systems do not encode proteins; thus population differences have no functional implications and reflect only the effects of population history.

The distribution of the 9-bp mtDNA deletion in South Indian populations provides further information about the history and evolution of this population. There is good evidence that this deletion has occurred independently in South Indian tribal populations (Watkins *et al.* 1999; Clark *et al.* 2000). The fact that the deletion is found primarily in tribal populations but has been observed in a small number of caste members underscores its potential for delineating gene flow between populations.

The comparison of between-caste genetic variation and genetic distances indicates little or no correlation between Y chromosome genetic distance and social rank. In contrast, there is a marked correlation between caste rank and mtDNA genetic distance (see Bamshad *et al.* 1998 for further details), and the correlation between caste rank and *Alu* distances reaches statistical significance.

It is possible that these results are due simply to mutation rate differences in different types of genetic systems, inadequate sample sizes, or other difficulties. These explanations, however, can be discounted (Bamshad *et al.* 1998). Instead, the explanation most consistent with the observed results (and with historical data) is that systematic female gene flow between castes has produced a correlation between mtDNA distances and social rank, while a lack of male gene flow has resulted in no correlation between Y chromosome distances and social rank. Estimates of inter-caste gene flow vary from 0.01% to 3% per generation (Balakrishnan 1982; Battacharyya 1993). Over the course of 100–150 generations since the founding of the caste system, this amount of gene flow could create the level of correlation observed here. On the other hand, the primary factors affecting Y chromosome variation in castes appear to be mutation, genetic drift, and the initial composition of the caste members (i.e. a presumably higher Caucasoid male contribution, as discussed above).

Conclusions

The results presented here shed light on the origins and affinities of south Indian caste populations. In particular, they are consistent with historical and archaeological data that indicate a primarily Asian origin for earlier south Indian populations, followed by an influx of Caucasoid migrants who established the caste system and tended to appropriate the highest positions in that system for themselves and their descendants. The Y chromosome and *Alu* distances support the hypothesis that many or most of these Caucasoid immigrants were males.

This study also demonstrates a correlation between mtDNA and *Alu* inter-caste distances and social rank but the absence of such a correlation for Y chromosome polymorphisms. Although this comparison needs to be made in additional caste populations elsewhere in India, its consistency with known historical patterns of inter-caste gene flow is encouraging. It implies that history can indeed leave a discernible imprint on the human DNA variation. Such imprints can in turn reveal much about our evolutionary past (see also Bertranpetit and Calafell, Chapter 5).

It is, therefore, important to realise that, not only in India, modern genetic techniques, in association with historical, social and demographical information from other sources, can be used to throw light on modern human populations and their past dynamics.

Acknowledgements

We wish to thank the individuals who participated in this study and members of the faculty and staff of Andhra University for assistance and helpful discussions. This research was supported by NSF grants SBR-9514733, SBR-9512178 and SBR-9818215, and by NIH grant GM-59290.

References

Balakrishnan, V. (1978). A preliminary study of genetic distances among some populations of the Indian sub-continent. *Journal of Human Evolution*, **7**, 67–75.

Balakrishnan, V. (1982). Admixture as an evolutionary force in populations of the Indian sub-continent. In *Human Genetics and Adaptation*, vol. 1, ed. K.C. Malhotra and A. Basu, pp. 103–145. New York: Plenum Press.

Bamshad, M., Fraley, A.E., Crawford, M.H., Cann, R.L., Busi, B.R., Naidu, J.M. and Jorde, L.B. (1996). MtDNA variation in caste populations of Andhra Pradesh, India. *Human Biology*, **68**, 1–28.

Bamshad, M.J., Watkins, W.S., Dixon, M.E., Rao, B.B., Naidu, J.M., Prasad, B.V.R., Reddy, P.G., Sung, S., Rasanayagam, A. *et al.* (1998). Female gene flow stratifies Hindu castes. *Nature*, **395**, 651–652.

Barnabas, S., Apte, R.V. and Suresh, C.G. (1996). Ancestry and interrelationships of the Indians and their relationship with other world populations: a study based on mitochondrial DNA polymorphisms. *Annals of Human Genetics*, **60**, 409–422.

Battacharyya, S. (1993). *Ecological Organisation of Indian Rural Populations*. Bangalore: Indian Institute of Science.

Bhattacharyya, N.P., Basu, P., Das, M., Pramanik, S., Banerjee, R., Roy, B., Roychoudhury, S. and Majumder, P.P. (1999). Negligible male gene flow across ethnic boundaries in India, revealed by analysis of Y-chromosomal DNA polymorphisms. *Genome Research*, **9**, 711–719.

Cavalli-Sforza, L.L., Menozzi, P. and Piazza, A. (1994). *The History and Geography of Human Genes*. Princeton: Princeton University Press.

Chandler, W.B. (1988). The jewel in the lotus. The Ethiopian presence in the Indus valley civilization. In *African Presence in Early Asia*, ed. I.V. Sertima and R. Rashidi. New Brunswick: Transaction.

Char, K.S.N., Lakshmi, P., Gopalam, K.B., Sastry, J.G. and Rao, P.R. (1989). Genetic differentiation among some endogamous populations of Andhra Pradesh, India. *American Journal of Physical Anthropology*, **78**, 421–429.

Char, K.S.N. and Rao, P.R. (1986). Glyoxylase I phenotypes in some endogamous populations of Andhra Pradesh, India. *Human Heredity*, **36**, 123–125.

Clark, V.J., Sivendren, S., Saha, N., Bentley, G.R., Aunger, R., Sirajuddin, S.M. and Stoneking, M. (2000). The 9-bp deletion between the mitochondrial lysine tRNA and COII genes in tribal populations of India. *Human Biology*, **72**, 273–285.

Das, K., Malhotra, K.C., Mukherjee, B.N., Walter, H., Majumder, P.P. and Papiha, S.S. (1996). Population structure and genetic differentiation among 16 tribal populations of central India. *Human Biology*, **68**, 679–705.

Felsenstein, J. (1993). PHYLIP (Phylogeny Inference Package) version 3.5c. Department of Genetics, University of Washington.

Jorde, L.B., Bamshad, M.J., Watkins, W.S., Zenger, R., Fraley, A.E., Krakowiak, P.A., Carpenter, K.D., Soodyall, H., Jenkins, T. *et al.* (1995). Origins and affinities of modern humans: a comparison of mitochondrial and nuclear genetic data. *American Journal of Human Genetics*, **57**, 523–538.

Jorde, L.B., Rogers, A.R., Bamshad, M., Watkins, W.S., Krakowiak, P., Sung, S., Kere, J. and Harpending, H.C. (1997). Microsatellite diversity and the demographic history of modern humans. *Proceedings of the National Academy of Sciences, USA*, **94**, 3100–3103.

Jorde, L.B., Watkins, W.S., Bamshad, M.J., Dixon, M.E., Ricker, C.E., Seielstad, M.T. and Batzer, M.A. (2000). The distribution of human genetic diversity: a comparison of mitochondrial, autosomal, and Y chromosome data. *American Journal of Human Genetics*, **66**, 979–988.

Karve, I. (1968). *Kinship Organization India*. Bombay: Asia Publishing House.

Kivisild, T., Bamshad, M.J., Kaldma, K., Metspalu, M., Metspalu, E., Reidla, M., Laos, S., Parik, J., Watkins, W.S. *et al.* (1999). Deep common ancestry of Indian and western-Eurasian mitochondrial DNA lineages. *Current Biology*, **9**, 1331–1334.

Krishnan, T. and Reddy, B.M. (1994). Geographical and ethnic variability of finger ridge-counts: biplots of male and female Indian samples. *Annals of Human Biology*, **21**, 155–169.

Labie, D., Srinivas, R., Dunda, O., Dode, C., Lapoumeroulie, C., Devi, V., Devi, S., Ramasami, K., Elion, J. *et al.* (1989). Haplotypes in tribal Indians bearing the sickle gene: evidence for the unicentric origin of ßs mutation and the unicentric origin of the tribal populations of India. *Human Biology*, **61**, 479–491.

Majumdar, D.N. (1958). *Races and Cultures of India*. Bombay: Asia Publishing House.

Majumder, P.P. (1998). People of India: biological diversity and affinities. *Evolutionary Anthropology*, **6**, 100–110.

Majumder, P.P. and Mukherjee, B.N. (1993). Genetic diversity and affinities among Indian populations: an overview. In *Human Population Genetics: A Centennial Tribute to J.B.S. Haldane*, ed. P.P. Majumder, pp. 255–275. New York: Plenum Press.

Majumder, P.P., Shankar, B.U., Basu, A., Malhotra, K.C., Gupta, R., Mukhopadhyay, B., Vijayakumar, M. and Roy, S.K. (1990). Anthropometric variation in India: a statistical appraisal. *Current Anthropology*, **31**, 94–103.

Majumder, P.P., Roy, B., Banerjee, S., Chakraborty, M., Dey, B., Mukherjee, N., Roy, M., Thakurta, P.G. and Sil, S.K. (1999). Human-specific insertion/deletion polymorphisms in Indian populations and their possible evolutionary implications. *European Journal of Human Genetics*, **7**, 435–446.

Malhotra, K.C., Chakraborthy, R., Bhanu, B.V. and Fulmali, P.M. (1980). Variation on dermal ridges in nine population groups of Maharasthra, India: intra- and inter-population diversity. *Human Heredity*, **30**, 307–315.

Maloney, C. (1974). *The Races in Peoples of South Asia*. New York: Holt, Rinehart and Winston.

Mantel, N. (1967). The detection of disease clustering and a generalized regression approach. *Cancer Research*, **27**, 209–220.

Mastana, S.S., Reddy, P.H., Das, M.K., Reddy, P. and Das, K. (2000). Molecular genetic diversity in 5 populations of Madhya Pradesh, India. *Human Biology*, **72**, 499–510.

Mountain, J.L., Hebert, J.M., Bhattacharyya, S., Underhill, P.A., Ottolenghi, C., Gadgil, M. and Cavalli-Sforza, L.L. (1995). Demographic history of India and mtDNA-sequence diversity. *American Journal of Human Genetics*, **56**, 979–992.

Murty, J.S., Muralidhar, B., Goud, J.D., Rao, P.J.M., Babu, B.R. and Rao, V.S. (1993). Hierarchical gene diversity and genetic structure of tribal populations of Andhra Pradesh, India. *American Journal of Physical Anthropology*, **90**, 169–183.

Nei, M. (1987). *Molecular Evolutionary Genetics*. New York: Columbia University Press.

Nei, M. and Roychoudhury, A.K. (1982). Genetic relationship and evolution of human races. *Evolutionary Biology*, **14**, 1–59.

Papiha, S.S., Singh, B.N., Lanchbury, J.S., Mastana, S.S. and Rao, Y.S. (1997). Genetic study of the tribal populations of Andhra Pradesh, South India. *Human Biology*, **69**, 171–199.

Passarino, G., Semino, O., Bernini, L.F. and Santachiara-Benerecetti, A.S. (1996a). Pre-Caucasoid and Caucasoid genetic features of the Indian population, revealed by mtDNA polymorphisms. *American Journal of Human Genetics*, **59**, 927–934.

Passarino, G., Semino, O., Modiano, G., Bernini, L.F. and Benerecetti, A.S.S. (1996b). mtDNA provides the first known marker distinguishing proto-Indians from the other Caucasoids; it probably predates the diversification between Indians and Orientals. *Annals of Human Biology*, **23**, 121–126.

Pitchappan, R.M., Balakrishnan, K., Sudarsen, V., Brahmajothi, V., Mahendran, V., Amalraj, S., Santhakumari, R., Vijayakumar, K., Sivalingam, P. *et al.* (1997). Sociobiology and HLA genetic polymorphism in hill tribes, the Irula of the Nilgiri Hills and the Malayali of the Shevroy Hills, South India. *Human Biology*, **69**, 59–74.

Quintana-Murci, L., Semino, O., Bandelt, H.J., Passarino, G., McElreavey, K. and Santachiara-Benerecetti, A.S. (1999a). Genetic evidence of an early exit of *Homo sapiens sapiens* from Africa through eastern Africa. *Natural Genetics*, **23**, 437–441.

Quintana-Murci, L., Semino, O., Poloni, E.S., Liu, A., Van Gijn, M., Passarino, G., Brega, A., Nasidze, I.S., Maccioni, L. *et al.* (1999b). Y-chromosome specific YCAII, DYS19 and YAP polymorphisms in human populations: a comparative study. *Annals of Human Genetics*, **63**, 153–166.

Roychoudhury, A.K. (1974). Gene differentiation among caste and linguistic populations of India. *Human Heredity*, **24**, 317–322.

Roychoudhury, A.K. (1977). Gene diversity in Indian populations. *Human Genetics*, **40**, 99–103.

Roychoudhury, A.K. (1982). Genetic relationships of Indian populations. In *Human Genetics and Adaptation*, ed. K.C. Malhotra and A. Basu, pp. 147–174. New York: Plenum Press.

Roychoudhury, A.K. (1984). Genetic relationship between Indian tribes and Australian aboriginals. *Human Heredity*, **34**, 314–320.

Saitou, N. and Nei, M. (1987). The neighbor-joining method: a new method for reconstructing phylogenetic trees. *Molecular Biology and Evolution*, **4**, 406–425.

Shriver M.D., Jin, L., Boerwinkle, E., Deka, R., Ferrell, R.E. and Chakraborty, R. (1995). A novel measure of genetic distance for highly polymorphic tandem repeat loci. *Molecular Biology and Evolution*, **12**, 914–920.

Singh, R.H. (1978). Dermatoglyphic variation in four castes of Uttar Pradesh, India. *Human Biology*, **50**, 251–260.

Sirajuddin, S.M., Duggirala, R. and Crawford, M.H. (1994). Population structure of the Chenchu and other South Indian tribal groups: relationships between genetic, anthropometric, dermatoglyphic, geographic, and linguistic distances. *Human Biology*, **66**, 865–884.

Smouse, P.E., Long, J.C. and Sokal, R.R. (1986). Multiple regression and correlation extensions of the Mantel test of matrix correspondence. *Systematic Zoology*, **35**, 627–632.

Tambia, S.J. (1973). The character of kinship. In *The Character of Kinship*, ed. J. Goody, pp. 191–229. Cambridge: Cambridge University Press.

Thangaraj, K., Ramana, G.V. and Singh, L. (1999). Y-chromosome and mitochondrial DNA polymorphisms in Indian populations. *Electrophoresis*, **20**, 1743–1747.

Thapar, R. (1980). India before and after the Mauryan empire. In *The Cambridge Encyclopaedia of Archaeology*, ed. A. Sherratt, pp. 257–261. New York: Crown.

Watkins, W.S., Bamshad, M., Dixon, M.E., Bhaskara Rao, B., Naidu, J.M., Reddy, P.G., Prasad, B.V., Das, P.K., Reddy, P.C. *et al.* (1999). Multiple origins of the mtDNA 9-bp deletion in populations of South India. *American Journal of Physical Anthropology*, **109**, 147–158.

Watkins, W.S., Ricker, C.E., Bamshad, M.J., Carroll, M.L., Nguyen, S.V., Batzer, M.A. and Jorde, L.B. (2001). Patterns of ancestral human diversity: an analysis of *Alu* insertion and restriction site polymorphisms. *American Journal of Human Genetics*, **68**, 738–752.

Appendix 8.1

Laboratory methods

Whole blood was collected in either 8 ml or 0.5 ml EDTA tubes by venous blood draw or finger prick, respectively. In most cases a hair sample was also collected. Samples were transported to the Laboratory of Biological Anthropology at Andhra University where DNA was extracted using a Puregene kit according to the manufacturer's specifications. These DNA samples were used to generate the data reported in this study: 411 bp of mitochondrial DNA (mtDNA) sequence from the hypervariable sequence 1 (HVS1) portion of the control region, presence or absence of a 9-bp deletion in the intergenic COII/tRNA^{Lys} region of the mtDNA genome, six polymorphic simple tandem repeat (STR) polymorphisms in the non-recombining portion of the Y chromosome (*DYS19, DYS288, DYS388, DYS389, DYS390,* and *DYS393*), three polymorphic Y-specific single-nucleotide polymorphisms (SNPs), and 40 autosomal *Alu* insertion polymorphisms.

To generate HVS1 sequence, a 1.1 kb amplicon from the mtDNA control region was amplified using conditions described above with primers UPL15996 and RPH408 and H16401 (Bamshad *et al.* 1998). Sequence for HVS1 was generated from the UPL15996 and H16401primer sites using ABI Dye-primer or dRhodamine sequencing reagents and an ABI 377 automated DNA sequencer. Sequence data were compared and edited using the Sequencher software package (Genecodes).

Multiplex genotyping of Y chromosome STRs was performed using an ABI 377 automated DNA sequencer. The PCR reaction contained one fluorescently end-labelled primer for each locus. DNA samples were amplified by PCR in 1X buffer (10mM Tris pH8.3, 50mM KCl) using 25 ng of genomic template DNA reaction product, 200 mM dNTPs, and 0.2 U Taq DNA polymerase complexed with *Taq*Start antibody (Clontech). Primer concentrations were optimised for each multiplex PCR panel. Thermal cycling was performed in a Perkin-Elmer 9600 PCR machine using a modified touchdown protocol where the first five cycles are done with annealing temperatures 2 degrees above the predicted average Tm followed by 25 additional cycles with annealing temperatures 4–6 degrees lower.

The Y chromosome genotyping utilised four-colour fluorescent dye chemistry; all six STRs were multiplexed in two PCR reactions and run in a single ABI lane. Products were resolved with urea denaturing polyacrylamide gels on the ABI sequencer using an internal size standard in each lane. Raw genotype data were collected using Genescan™ software (ABI), and gel files were analysed using the Genotyper software package (ABI).

The 40 polymorphic *Alu* loci were genotyped by amplifying 25 ng of genomic DNA in a standard 30-cycle, 3-step PCR. Appropriate annealing temperatures and additives were optimised for each system. For most systems, samples were amplified with 5 µl of cresol red loading buffer (34% sucrose, 0.02% cresol red) to eliminate the need to add dye to the samples before gel loading. Following PCR, the samples were loaded on multiple-combed 3% Nusieve agarose (3 : 1) gels and electrophoresed at 175 volts for 2 hours. Ethidium bromide stained gels were visualised by UV and documented.

Statistical methods

The data set consists of 411 bp of HVS1 sequence for 640 individuals who were also typed for the 9-bp mtDNA deletion. In addition to the caste members described above, the analysis included 143 Africans, 61 Asians and 120 Europeans (Jorde *et al.* 1995, 1997, 2000). For mtDNA sequence data, gene diversity was estimated for each population as $\pi = (n/n - 1) \, \Sigma x_i x_j d_{ij}$ where n is the number of DNA sequences examined, x_i and x_j are the population frequencies of the ith and jth type of DNA sequences, and d_{ij} is the proportion of nucleotides which differ between the ith and jth types of DNA sequence. For the Y chromosome STRs and *Alu* polymorphisms, gene diversity was estimated as $1 - \Sigma x_i^2$, where x_i is the estimated frequency of the ith allele in the system. Gene diversity levels within and between populations were used to estimate the proportion of genetic variance due to subdivision (G_{ST}). For mtDNA data, genetic distances between individuals were estimated using the Kimura 2-parameter model assuming a transition/transversion ratio of 10 : 1. For Y chromosome STRs, a measure was used in which the distance is weighted by the difference in allele size [Shriver *et al.*, 1995 #3179]. Nei's (1987) standard distance metric was used to assess genetic distances for the *Alu* polymorphisms. Networks were assembled from the distance matrices using the neighbour-joining method (Saitou and Nei 1987) implemented in the PHYLIP package (Felsenstein 1993).

To assess the correlation between inter-individual caste rank differences and inter-individual genetic distances, two $n \times n$ matrices were formed, where n is the number of individuals. For the first matrix, inter-individual distances were based on the proportion of *Alu* insertion/deletions shared by each pair of individuals. To form the second matrix, each individual was assigned a score according to his rank in the caste hierarchy (i.e. Brahmin = 1, Kshatriya = 2, Vysya = 3, Kapu = 4, Yadava = 5, Relli = 6, Mala = 7, Madiga = 8). An inter-individual matrix of score distances was formed by comparing the absolute value of the difference between the scores of each pair of individuals. The correlation between the two distance matrices was assessed using the Mantel matrix comparison test (Mantel 1967; Smouse *et al.* 1986). This test employs a permutation technique to assess the statistical significance of correlations between pairs of matrices.

9

Fertility, mortality and migration transitions in association with socioeconomic modernisation among highland minority populations in Southeast Asia

PETER KUNSTADTER

Introduction

This chapter compares causes and consequences of population growth in the Lua' and the Hmong highland minority groups in Southeast Asia. The study illustrates roles of cultural patterns in, and impacts of, modernisation on population dynamics. It exemplifies many of the processes introduced in earlier chapters in relation to demography, change in social and health conditions, and population genetics. It relates changes in mortality, fertility and migration, and in geographic and social boundaries, to socioeconomic modernisation, including public health programmes, education, development and the market economy. This condensation of more detailed papers by Kunstadter and colleagues (see References section) suggests the value of holistic and longitudinal approaches.

Data were collected over four decades by conventional anthropological participant observation, health examinations, unstructured and structured interviews, surveys and questionnaires, some modelled on those developed for international comparative studies, such as the *Demographic and Health Survey*, with appropriate modifications based on discussions with members of the populations. Native speakers conducted most surveys.

Determinants of primary population variables

In Lua' and Hmong populations almost all fertility is confined to marriage. Thus age at marriage, proportion who marry, and divorce rates are important determinants of fertility, as are use of deliberate methods to delay, space or end pregnancies (Davis and Blake 1956).

Important biological determinants of fertility include age of reproductive maturity and senescence. These are influenced by nutritional status and infections, especially sexually transmitted diseases. Breastfeeding suppresses ovulation and protects infants from infectious diseases. Death of breastfeeding infants may increase fertility if ovulation returns to the mother more quickly. Fertility and mortality variables, discussed by Hinde (Chapter 2), are described in greater detail in this chapter.

Definitions and measurements of migration differ and complicate the understanding of this variable. Different measures of migration include more or less permanent change of residence associated with marriage, change of residence between time of birth and time of census or survey, and temporary change of residence. Place of residence, and thus the geographical definition of a population, is socially bounded and often has important legal implications. *De jure* (legal) residence and *de facto* (actual place of) residence may not coincide. In residentially stable populations such as Lua' in traditional times, places of residence and work were identical and there was no difference between *de jure* and *de facto* residence. In recent years the volume of migration has increased, the timing and purposes of migration have become more varied, the legal implications of place of residence have proliferated, and migration has become a primary determinant of population distribution (see Clarke, Chapter 3). The *de jure–de facto* distinction has become more important. Daily commuting to work or school further complicates the problem of definition of geographical and temporal boundaries of populations.

Population boundaries are also subject to arbitrary choice of criteria for definition. In isolated populations geographic, genetic, ethnic, cultural, social and economic boundaries coincide. With loss of isolation in modern societies and economies, these criteria often vary independently, and populations defined by one criterion overlap with populations defined by other criteria (see Introductory chapter). Mobility expands the economic, demographic and genetic boundaries of populations.

Traditional Lua' and Hmong models

Traditionally both Lua' and Hmong were patrilocal and patrilineal. Both were traditionally swidden farmers, who cut and burned forest for upland cultivation of rice, but their systems of land tenure and land use were very different and were crucial in determining responses to population growth and distribution. Traditional boundaries of Lua' economy, and population for the most part, coincided with the village, while Hmong economy and society extended far beyond village boundaries.

Lua' model – conscientious conservationism

At present about 10,000 Lua' live in highland villages in Thailand, speaking several closely related dialects of a Mon-Khmer language. Lua' were the autochthonous population of northern Thailand and defined themselves in terms of their cultural distinctiveness. They took pride in serving more spirits than any other ethnic group, in their language, in their ability to learn other languages, and in their attachment to a specific village, its land and its spirits. They lived in villages that had been settled within the same territory for hundreds of years and had a 'royal' lineage of chief priests (*samang*) and leaders. Commoners were obliged to pay tribute to the *samang*, whose lineage traced its descent from a Lua' king hundreds of years ago. Before the end of the nineteenth century, when modern Thailand was consolidated under the rule of the Bangkok kings, Northern Thai princes granted rights to land and for self-government to these villages. In return villages paid princes small amounts of forest products as tribute (Nimmenahaeminda 1965).

Lua' practised a conservative method of subsistence farming. All villagers farmed within the bounded village territory and cut, burned and cultivated their fields for a single year in a coherent block of fields. Lua' farmers felt strongly that their fields should be bordered by the fields of others from their same village, or by the edge of uncut forest, not by fields of non-Lua'. When other ethnic groups expanded into traditional Lua' territory and non-Lua' farmers cut swiddens within Lua' land, Lua' villagers sometimes withdrew from their traditional swiddens so as to have only Lua' neighbours. They used hand tools to cut and scrape weeds, disturbed only the top few centimetres of soil, and left tree stumps in fields to allow trees to regrow. Natural vegetation restored forest cover and soil fertility during the nine years before farmers returned to cut the same fields again (Kunstadter 1978; Sabhasri 1978; Zinke *et al.* 1978). The main subsistence crop was upland rice, supplemented by a wide variety of food crops, cotton for homespun fabric, etc. Lua' had no cash crop but occasionally sold rice when they had surplus or forest products if they needed cash.

Village religious leaders could fine villagers who violated restrictions on land use such as cutting trees in a sacred forest or burning fields outside the times when all fields were burned simultaneously. Because village land was bounded and there were strong feelings that villagers should farm only within the village land, this system could continue only if the village population did not increase and villagers agreed on the rules. Every household had to contribute to costs of communal ceremonies and supply labour for communal tasks, especially clearing firebreaks around the block of fields to be burned and fighting forest fires.

Migration into villages was rare because access to land was by inheritance. In-migrants were usually brides from nearby Lua' villages. People moved out of the village temporarily in case of economic disaster such as a fire that burned stored rice. Permanent out-migration occurred rarely, when a woman married outside the village, or when a household was chronically unable to produce enough food or keep up its obligations to village-wide ceremonies, or if they violated important taboos and thus threatened the village with the wrath of vengeful spirits.

Nuclear or patrilineal stem families, with husband, wife, unmarried children, one married son and his wife and children, averaged under six members. Households with more than seven or eight members were unusual (e.g. Kunstadter 1984a). When a younger brother married, his older married brother established a new nuclear household. Household members, often assisted by a small number of relatives and friends, conducted farming activities such as clearing and weeding. Households assembled large crews of workers for time-limited activities, such as planting and harvesting, through traditional obligations to kinsmen and reciprocal labour obligations, and by providing a meal containing meat from a sacrificed animal. Households transmitted land use rights to heirs who were usually sons. If there were no sons, a co-resident son-in-law or daughters or wives could inherit. Population growth was slow, and on average each household had only one male descendant, and use rights were relatively stable over time. The chief priest could reallocate use rights between households if major inequalities developed in access to land.

Lua' customs delayed age at marriage, thereby reducing potential fertility in several ways:

- Bridegrooms paid a substantial bride price to their bride's family when they married. They spent several years accumulating bride price by selling surplus subsistence crops, pigs or forest products, or by wage work outside the village before they could marry. Lua' felt that spouses should be of similar age. This meant that women's age at marriage was constrained by economic requisites for men's marriage.
- Lua' had a strong feeling, backed by religious sanctions and payment of a fine, that older siblings should marry before younger siblings.
- There was no polygyny. Thus any 'surplus' women could not marry. There was a preference for cross cousin marriage. A man who did not marry his cross cousin had to pay her family a fine.
- Lua' preferred village-endogamous marriage but marrying within one's own patrilineal extended kin-group of about four generations' depth was taboo. In small villages this meant that there were often few potential mates. If the number of suitable mates was

too small, village leaders occasionally reduced the number of generations of genealogically defined exogamy (Kunstadter 1966). Bachelors also could seek brides in nearby Lua' villages.

- Girls who became pregnant before marriage were required to marry but this was a relatively rare event because religious sanctions and gossip backed prohibitions on sexual intercourse before marriage.
- Divorce was rare, but did not require repayment of bride price.

These constraints delayed age at marriage for both young men and young women, and thus decreased the potential number of children any woman could bear.

Lua' parents wanted children to help work their fields and to take care of them in old age and traditionally believed that children should care for the spirits of their ancestors. Sons normally took on the burden of care for the ancestral spirits. If a couple had no sons, an uxorilocally married son-in-law took on the obligation, along with access to the household's land use rights. Although Lua' parents wanted sons to carry on the family line, and to assure that their own spirits would be fed after they died, they were not nearly as anxious to have sons as were Hmong who do not allow matrilocal marriage and inheritance through a daughter's husband. Because Lua' were subsistence farmers without a cash crop and without a convenient way to store any surplus they accumulated, and because leadership positions in the community were determined by inheritance rather than achievement, there was no strong economic or political motive for having large numbers of children.

Hmong model: exploitation of resources on an expanding frontier

The majority of the Hmong population (over 8 million) lives in southwestern China, where the ethnic group originated (Geddes 1976, Lemoine 1972, Yang Dao 1992). Another half million live in northern Vietnam, a quarter million live in Laos, and over 100,000 live in Thailand in more than 200 villages. Hmong moved from China because of persecution by ethnic Chinese, but the main reason for migration within Southeast Asia was local exhaustion of resources and population growth. Hmong began moving into northern Thailand from Laos and China in the 1890s at about the time the Bangkok government took political control of northern Thailand from the ethnic Northern Thai princes. Hmong had no feudal relationships with the princes. Hmong households usually formed villages in unoccupied territory at high elevations where they could convert forests into fields. Villages were temporary agglomerations of households, not permanently organised communities. Hmong villages did not have inherited positions of religious or civil leadership, and village leaders did not control access to or use of land.

Hmong traditionally grew labour-intensive upland rice as their main subsistence crop and opium as a cash crop and for home consumption. After cutting, burning and planting their fields they used hoes for deep, clean cultivation and dug out tree stumps, thereby limiting the possibility of forest regrowth. They planted the same field repeatedly, but abandoned it when soil fertility was exhausted. Individual households cleared and farmed nearby unused land in the same way. If unclaimed land was too distant from the village, households moved to new locations where land was available (Keen 1978). Unlike Lua', Hmong were willing to move to areas surrounded by other ethnic groups, or used lineal or collateral kinship connections to gain access to land in distant villages.

Unlike Lua' traditional religion, Hmong religion is not localised or transmitted by inheritance of land. Important Hmong ceremonies are related to ancestor worship, weddings, funerals and diagnosis and treatment of illness, but not to spirits of the place. Spirit doctors (shamans) are recruited by being 'called' by spirits, not by inheritance. Hmong shamans do not control access to or use of land.

The Hmong population comprises a number of non-localised exogamous patrilineal descent groups (clans), with strict taboos against marriage or sexual relations between members of the same clan. Hmong identify themselves in terms of their clan, extended family, language and dialect, and only secondarily by region of origin, not specific village. Strangers who belong to the same clan but cannot immediately identify a common ancestor ask each other for details of ancestor worship and other rituals to determine how closely related they are (they presume that the greater the differences, the more generations have passed since descent from a common ancestor). Clans provide assistance to their members in times of need, such as a place for a traveller to stay in the home of a fellow clan member. Clan members are concerned about the spiritual condition and secular reputation of their kin group, and support one another in conflicts with members of other clans. Women join their husband's clan when they marry.

Patrilineally extended multigenerational Hmong households often include a father, his wife or wives, their unmarried children, their married sons, sons' wives and children, and sometimes their sons' sons and their wives and children. Household sizes average over 10 persons and range up to 100 or more individuals living in a common house and collaborating economically (Kunstadter 1984a). Leadership within extended families is based primarily on ability, often judged by accumulated wealth, not necessarily on seniority. The leader of an extended family from the largest clan often assumes leadership of a village. This man is often a strong and respected leader, but his power is not backed by religious sanctions over people outside his extended family. He has little control over access to land or agricultural activities, which are matters of individual initiative.

Nuclear families within the household cultivate major subsistence crops co-operatively and pool their harvest, but individual families usually cultivate their own cash crops and retain profits or pool them for some household purpose such as bride price. Closely related households often live in nearby houses, work their fields separately, and collaborate for religious ceremonies or for planting or harvesting when a large labour force is needed temporarily. Children participate in activities such as weeding from an early age. Marriage is often village-exogamous. New Year celebrations are one occasion for visiting between villages and provide opportunities for unmarried boys and girls to meet. Young men also sometimes travel in a group to another village for courtship.

Hmong have a number of customs that they manipulate to *reduce* constraints on marriage and encourage marriage (especially of women) at a young age (Kunstadter 2002).

- Although a Hmong groom's family must pay a substantial bride price to the family of the bride, bride price can be paid over time, or the young man may substitute it with temporary work in the bride's house. To reduce economic constraints to marriage, Hmong leaders in several northern Thailand provinces set an upper limit on the price that a bride's family could demand when bride prices were increasing in the 1980s.
- Preference for older siblings to marry before younger siblings is not enforced.
- Husbands are generally older than their wives. There is no strong feeling concerning relative age of spouses, but parents and grooms expect that girls should marry while young, sometimes even before they are physically mature.
- Because the Hmong population has expanded rapidly, each successive cohort is larger than its predecessor. The male/female difference of about two years in median age of marriage results in more women than men of marriageable age allowing near-universal marriage for men and frequent polygyny.
- Hmong men marry polygynously if their first wife does not produce a male heir, or in order to increase their household labour supply, or for sexual desire. Men who take second or third wives often marry younger women. This helps to ensure that every woman marries.
- Unmarried girls who become pregnant are required to marry. This is more common than among the Lua' because prohibitions on sexual intercourse before marriage are not rigidly enforced. Parents fear disgrace if their unmarried daughter becomes pregnant, but also fear gossip if their unmarried daughter becomes 'too old'.
- Parents who know each other or are already related by marriage sometimes make advance betrothals of their children (e.g. cross cousin marriage).
- Hmong traditionally recognised marriage by abduction or 'bride capture', but have largely abandoned this practice in recent years because Thais view it as 'uncivilised'. Although this form of marriage was appropriate when Hmong villages were isolated and it was hard to find a bride, nowadays with pickup trucks people can

go anywhere and there is no need to retain this custom that leads to disputes between in-laws.

- Hmong traditionally practised levirate marriage of younger brother with older brother's widow to assure that the household would retain her labour and her children.

- If the marriage results from the wishes of the young couple, parents of bride and groom should agree on the bride price before the marriage. In fact, most such couples elope. Because Hmong consider that a couple that has spent a night together has married, eloping eliminates the ability of the bride's family to forbid the marriage and greatly reduces their bargaining power in negotiating the bride price.

- Divorce is strongly discouraged because unless the husband is notoriously cruel to his wife, the wife's parents must repay the bride price if the wife leaves her husband. Most parents are unable and unwilling to make such a repayment so often the only recourse a woman has in an unhappy marriage is suicide. Also, once a woman marries, she becomes a member of her husband's clan and thus is not supposed to spend the night in her parents' house in the presence of her father's ancestral spirits. If she returns home, her parents have to build her a separate room with a separate entry door.

These customs encourage marriage at a relatively young age (median age of 17 for women), and permanence of marriage. Almost all Hmong women marry young and are married for most of the time when they are physiologically capable of pregnancy. This allows high fertility.

Large family size is an asset in disputes. Because the opium cash crop allowed Hmong to convert household labour to a store of value (usually silver bars), the advantage of increasing household or family labour force was a strong motive for fertility. Hmong believe they must honour their ancestor spirits in order to ensure their own success. Only sons, not sons-in-law, can worship the household's patrilineal ancestors and increase the household labour supply by bringing in wives and having children. Uxorilocal marriage or inheritance through females is not allowed. Only sons kill cows or other large animals for their parent's spirit when the parent dies. Thus parents have a strong desire for sons, and also want an equal number of daughters in order to collect bride price when their girls marry.

Wives also have strong motives for high fertility. A wife who does not produce a son fears losing her husband to a second wife, and also needs a son to support her in old age, especially if she loses favour with her husband. Women also want daughters for companionship and to help with household chores, at least until a son brings in a wife. Wives want daughters-in-law, because they raise the mother-in-law's status by replacing them at the bottom of the household pecking order.

Demographic differences associated with the different models

In the early 1980s mean Hmong household size (12.6 persons) was double that of Lua' (5.9 persons), crude birth rate was almost double (49.8/1000 compared with 29.4/1000), crude death rate was half (8.9/1000 cf. 19.2/1000), death rates, standardised to the age distribution of a nearby lowland town, was about one quarter (3.7/1000 compared with 13.7/1000), and the Hmong rate of natural increase was four times that of Lua' (40.9/1000 compared with 10.2/1000) (Kunstadter 1986). Surveys on larger Hmong populations showed similar rates (e.g. Kunstadter and Kunstadter 1990; for other demographic data see Kunstadter 1971, 1983a, 1984b; Kunstadter *et al.* 1989).

Major external changes affecting population dynamics

Since 1961, National Social and Economic Development Plans have guided the economic development policies of the Royal Thai Government (RTG). Earlier plans emphasised the infrastructure (e.g. dams, roads, primary education, public health) and the growth of industry in lowland urban centres. More recent plans stress 'human capital' (e.g. secondary and higher education) and reduction of economic disparities (Sukwong 1999; Anon. 2000). The earliest major alteration in vital rates for Lua' and Hmong highlanders was a decline in mortality associated with control of major infectious and parasitic diseases. Smallpox was eliminated in the late 1940s and malaria was controlled beginning in the late 1950s. Malaria control may have been more important for Lua', who live at altitudes where malarial mosquitoes breed, than for Hmong, who usually live at a higher altitude in Thailand. Control of malaria was important for Hmong in increasing their direct participation in the market economy, because they 'always' got malaria when they went to market in the lowlands. Road building by the RTG allowed access to markets, added sources of income, and evened out seasonal and periodic fluctuations in food supply of subsistence farming for highlanders. Reduction of fertility lagged behind reduction in mortality, and as a result highland populations grew rapidly.

For Hmong, but not Lua', other important factors were government relocation projects and consolidation of communities associated with the communist insurgency in the late 1960s to 1980s, and suppression of opium cultivation in the late 1980s. Hmong in Thailand, but not Lua', were affected by the Indochina wars, because of the flight of tens of thousands of Hmong refugees from Laos to Thailand and their subsequent resettlement in the United States, France and elsewhere after the end of the war in 1975. Other factors affecting both Lua' and Hmong include globalisation of Thailand's agricultural economy for imports and exports, mass media communication, modernisation of lifestyles, consumerism, control of land use in the highlands, education and, especially for Lua', missionary activity.

Lua' responses to population growth

Pa Pae villagers followed several strategies in response to population growth.

- Farmers intensified agriculture (cf. Boserup 1965) by carving terraced, irrigated rice fields out of the hillsides and cultivating them annually, while swidden fields in the traditional rotation system could be cultivated in only one year out of ten. The creation of irrigated fields up the hillsides accelerated in the 1960s and 1970s,

and for a while rice production kept pace with population growth
(Kunstadter 1985). Eventually this expansion was limited by the
amounts of suitable land, labour and water.

- They shortened fallow time of swidden fields from nine to eight,
and sometimes seven years, with consequent loss of soil fertility.
Yields declined, insect pests caused crop failures, and economic
problems increased in the 1990s.

- In the late 1960s and early 1970s development project workers
persuaded Pa Pae farmers to grow coffee and soybeans as cash
crops. These efforts were economically unsuccessful for technical
reasons. Eventually villagers were successful in growing *shiitake*
mushrooms as a cash crop. In Lua' villages with better road access
to market, farmers grew cabbages as a cash crop.

- In the early 1960s increasing numbers converted to Christianity
(Kunstadter 1983b). Traditionally, individuals or households that
could not afford to contribute to the cost of communal rituals
moved or were forced out and became members of the ethnic
group of the community into which they moved. Several Pa Pae
households converted to Christianity in the 1960s. Christians
stayed in the village and continued to use village lands, but did not
have to sacrifice animals to treat illness, to assure the harvest, or
for communal well-being. This lowered costs for Christian households and allowed population growth despite low agricultural production. By 1998 half of the village households were Christian.
Either Christians married Christians or animist spouses converted
to Christianity. Because the pool of eligible Lua' mates who were
Christian and not members of the same kin group was small,
marriage with non-Lua' Christians was encouraged. Two-thirds
of the non-Lua' in-migrant spouses were Christians. This strategy may have benefited individual households, but it increased
the population relying on village-based resources. The effect on
village resources was smaller than it might have been because
many of the in-migrants eventually moved out.

- They reduced fertility by family planning. The proportion of married women of reproductive age who were using modern family
planning increased from none in the 1960s to about 65% by the
early 1990s.

- They migrated to the lowlands for employment. In 1968 the *de facto* village population was 226 (Kunstadter 1972), and our 1998 survey showed it had increased to 305, while the number of former residents who moved elsewhere increased threefold, from 34 to 137, so that much of the population increase in the 30 intervening years was compensated by out-migration. In 1968 the most popular destination was the nearby district town, about 20 km away. By 1998, 88 former residents had moved to the city of Chiang Mai, 150 km away, compared with none who lived there in 1968. All of those who moved out were under the age of 45. The proportion of women among the out-migrants remained about the same (53% in 1968, 54% in 1998). However, part of the increase in village population was due to migration into the village. The proportion of in-migrants increased from 3.5% in 1968, to 11.5% in 1998. The age distribution of village residents in 1998 showed a major deficit among those aged 15 to 44, while the distribution of out-migrants showed a surplus in these ages. Age distributions in 1968 and 1998 suggest a decline in fertility.
- Consistent with the massive out-migration from rural places and agriculture in general to urban settings and economic modernisation (see Clarke, Chapter 3), parents recognised the values of education and Thai language skills and sent both sons and daughters to school.

Hmong responses to population growth

The Hmong population was growing at a rate of about 4% per year in the early 1980s because of a surplus of births over deaths.

In the mid- to late 1980s the Royal Thai Government began restricting cultivation of opium and curtailing access to forest land. Hmong villagers responded in a number of ways.

- Hmong intensified and diversified agriculture with a great variety of non-narcotic cash crops (Kunstadter 2000). Most of the new cash crops required heavy capital investment for machinery and supplies. Accumulated capital from opium probably made it easier for Hmong than for Lua' to make the transition from land-extensive, labour-intensive farming to land- and capital-intensive farming, a transition facilitated by Hmong familiarity with the market economy. Tractors and herbicides reduced the need for labour, but large amounts of labour are required for brief periods such as harvest. Hmong farmers traditionally used household labour and reciprocal exchange between related households, but recently they have begun to use non-Hmong labour, including illegal immigrants from as far away as Bangladesh.

- By 1987 an average of about 25% of women of reproductive age were using modern family planning; the rate was higher in communities closest to market and reached 75% among Hmong living in urban places.

- In 1987, parents in Hmong communities, where land scarcity and government pressure was greatest, already perceived that the economic future of farming was limited and that land resources would be inadequate for their children. This increased their aspiration for education and for non-farm occupations for their children (Kunstadter *et al.* 1991, 1994).

- By the mid-1990s many Hmong farm families had stopped producing subsistence crops and, unlike Lua', were buying most of their subsistence products. Hmong made the initial rapid change of agriculture from subsistence plus opium to a wide variety of cash crops without a major decline in economic condition. Perhaps as a result, conversion to Christianity has not been an important response of Hmong, and there has been no change in traditional marriage and post-marital residence patterns.

- The traditional reason for migration was search for land. Comparison of reasons for migration of village subgroups before and after 1964 shows major increases in migration related to RTG policies. Before 1964, out of 685 recorded moves of village subgroups, 84% were related to access to land and farming, compared with 59% of 414 moves thereafter; before 1964, only 3.6% of moves were related to pressure or resettlement by the RTG compared with 24% thereafter. Before 1964, fewer than 1% were related to lack of development or infrastructure, compared with 6% thereafter.

- Distances, volume and purposes of migration increased with new road networks, motor vehicles, and electronic communication. Some Hmong farmers now live hundreds of kilometres from their registered (*de jure*) home during farming seasons. Some Hmong women now travel hundreds of kilometres for trade, unaccompanied by men. Hmong women now stay for long periods in Chiang Mai where they sell flowers, while their husbands commute to the home village to grow and transport flowers to market. Resettled Hmong communities are much larger than traditional communities, and have better roads, schools and health facilities, and better access to market. Hmong merchants travel and engage in international trade with Hmong in other countries (China, Laos, Vietnam, USA, France), Hmong scholars and refugees living in the West travel to Laos, Thailand and China. A few Hmong from Thailand study in the USA and Hmong from China study in Thailand.
- In the 1990s Hmong moved to urban places in rapidly increasing numbers. For example, in Chiang Mai there has been a rapid increase in numbers of students, workers in factories and stores and those self-employed as merchants, as well as Hmong drug addicts exiled from their home villages.

Effects of exogenous changes on population and ethnic boundaries

Both populations grew as a result of control of diseases due to public health interventions and improved access to food supplies.

Ethnic boundaries are defined both inside and outside the group in question (see the Introduction to this volume). Both Lua' and Hmong traditionally defined themselves by their own language and culture. 'Culture' for the Lua' traditionally included religious practices that bound them to particular localities, while the Hmong were bound to their patrilineal ancestors and clan relatives, and were characteristically mobile. The Hmong cash economy connected them with national and international markets, and their social organisation, at least theoretically, connected all members of the same clan wherever they might be located. Thus traditional Hmong self-identity is inherently more portable than Lua' identity. Another difference between Lua' and Hmong adaptations is that while urban migrants from a single Lua' village have formed informal co-operative groups, Hmong have established formal supra-communal organisations based in urban Chiang Mai. Hmong in the USA have formed numerous ethnic-based formal organisations along clan lines (e.g. Fresno Yang Clan Association), around special interests that cut across village, clan and region of origin. Supra-village Hmong organisations are also reported in China (e.g. Kweichou Hmong Association).

Boundaries of the economic system have expanded for both groups as they now produce crops and other products for national and international markets. Breeding population boundaries have been modified for Lua' with the in-marriage of non-Lua', while ethnic endogamy has been maintained by Hmong in Thailand. In the USA, ethnic endogamy may be breaking down at least for educated women seeking to escape from male-dominated Hmong society. Some refugee Hmong men travel to Asia seeking wives who they believe will be more naïve and easier to manage than Hmong women who are familiar with Western divorce laws. Modernisation of transportation and communication have made it easier for both Lua' and Hmong to maintain connections between migrants and their communities of origin. At the same time, increased exposure to members of other ethnic groups, either directly or through mass media, has made both migrants and non-migrants aware of different cultural standards.

The blurring of cultural boundaries between populations that were once distinct is indicated, for example, by acceptance of some of the rhetoric of Thai family planning programmes or by Hmong modifications of marriage customs, such as abandonment of bride capture. Parents in both groups now do not want their children to suffer the hardships of traditional highland agriculture, and recognise the need for Thai language education and migration into a Thai-dominated environment for jobs that yield a salary.

There are profound differences between Lua' and Hmong with regard to their current definitions of ethnic identity. As a result of the international diaspora of Hmong refugees there are now substantial populations of Hmong in the West. Many refugees passed through

Thailand on their way to third countries and many established and retained contact with relatives in Thailand. Many Hmong from Laos who declined to be sent to third countries remain in Thailand, and some have moved, legally or otherwise, into highland Thai Hmong communities. Education in the West, especially of Hmong academics, leads to contacts between Hmong in China, Thailand, Vietnam, the USA, France and Australia, and broadens recognition of the extent of the Hmong population. International contacts between Hmong are exploited commercially, for example the production and international sale of Hmong language video tapes. Hmong leaders in the West now cultivate Hmong ethnicity, both out of genuine ethnic pride and for political reasons. Hmong are now members of an ethnic group with a world-wide distribution that has much more consciousness of its identity than did people living in remote villages 25 years ago. International travel strengthens ties between widely dispersed people, and the geographically dispersed clan system facilitates communication and feelings of kinship among people who may be only very distantly related biologically.

By contrast, Lua' are still a small population without any geographically dispersed social organisation. Consistent with a well established pattern in Southeast Asia (Leach 1954), many Lua' villages in the Chiang Mai valley have 'become Thai', by speaking the local Thai dialect and practising Buddhism (with only occasional use of traditional Lua' rituals). In the past individual Lua' moving from the mountains to the lowlands sometimes 'passed' as Northern Thai by speaking Thai, but as the number of Lua' living in more or less coherent communities increased they tended to retain their Lua' identity, even if they became Buddhists and no longer spoke Lua'. Because of their investment in Lua' language Bible translation and literacy in Lua', and the general identification of Thailand and 'being Thai' with Buddhism, Lua' Christians and missionaries have an interest in maintaining Lua' language and identity. Also, conversion from traditional animism to Christianity sometimes blurs ethnicity for Lua' Christians by establishing a bond with members of other previously distinct ethnic groups. This means that Christian Lua' migrants to urban areas may call on multiple ethnic identities (Lua' ± 'hilltribe' ± Thai ± Christian) under various circumstances.

Consequences for population dynamics of ethnic labelling of minorities by the dominant majority

The attitude of the dominant majority and the RTG towards highland peoples has been an important agent of ethnic change for both the Lua' and the Hmong. The official policy toward highland minorities is one of assimilation, with use of the Thai language in schools and the imposition of Thai-sounding last names. Hmong, whose family names refer to their clans, fear that loss of their clan names will lead to culturally defined incest as a result of inadvertent violations of clan exogamy.

Lowland Thais generally group all highlanders, regardless of important linguistic, cultural and historical differences, together as *chao khao* (roughly equivalent to hillbilly) as opposed to *chao rao* (our people). Lua', because of their long history within what is now the Kingdom of Thailand, have been considered as citizens, but many Hmong, because they are relatively recent arrivals in the Kingdom, and for other historical reasons, have not been considered citizens, regardless of where they were born. Hmong are also more often than Lua' the subject of negative stereotypes (e.g. Khongsanit 2001), as in the 1970s when the term *Maeo Daeng Kommunit* (red communist Hmong) was often applied indiscriminately to all highlanders. Many government and quasi-government development programmes are run as if there were no differences between highland ethnic groups. Agricultural project workers, teachers or health workers from one ethnic group are assigned to work with people from other ethnic groups whose language and culture they do not know. Common treatment and designation by Thai society has led to some degree of acceptance of the designation of 'hilltribe' by all highlanders. Many highlander children have been eligible to attend welfare boarding schools in the lowlands, which include members of other ethno-linguistic groups. Students learn to speak a common language (Thai) and establish friendships across ethnic groups that traditionally would never have occurred.

Because they are 'exotic' and their traditional clothing is colourful, highlanders have been used to entertain official visitors, and from the 1970s used as tourist attractions. The pooling of members of diverse ethnic groups in commercial 'cultural centres' has also led to the formation of friendships between members of diverse ethnic groups. Foreign governmental and non-governmental organisations often consider highlanders to be members of a single class for development projects. The Royal Forest Department has made statements and implemented regulations applying to all highlanders and has attempted to force people off the land, regardless of their pattern of use of highland area or the length of time they have occupied the area or used the land (Banijbatana 1978; Ratanakorn 1978; Sukwong 1999: 127–128). Common treatment by RTG officials has given diverse ethnic groups common interests in protecting their traditional land use systems whereas these groups have been and are in competition for land and water. Probably unintentionally, this common treatment is creating a coalescence of people of diverse ethnicities and establishing a consciousness among highlanders of an identity as members of a minority, often negatively valued. This sometimes surfaces in the rhetoric of Thai politicians. Thus members of diverse highland ethnic groups joined together in political action in 1999 in a protest in front of the Chiang Mai Provincial Offices regarding access to land and citizenship. This appears to be the first public demonstration of class-consciousness that cut across traditional ethnic boundaries.

In summary, the common treatment by Thai officials is creating a greater coalescence of people of diverse traditional ethnicities.

Conclusions

Comparison of culturally different populations that are exposed to similar exogenous forces in similar environments assists in understanding the role of cultural factors in population dynamics. The Lua' and the Hmong traditionally had major differences in fertility, mortality and migration rates, in marriage patterns, in traditional behaviours regulating relationships between population and resources, and in mechanisms for maintaining ethnic boundaries. Both groups have undergone rapid population increase as a result of public health efforts, and both now have declining birth rates associated with use of modern family planning methods.

Government policies and the attitudes of the dominant majority population have been similar for these culturally different minority populations, resulting in the blurring and redefinition of some ethnically defined boundaries from the point of view of both the minorities and the majority. However, there are major differences in Lua' and Hmong responses to recent rapid increases in population size and controls imposed on use of land resources. Compared with Lua', whose attempts at agricultural intensification have been relatively unsuccessful, the relative success of Hmong in agricultural intensification may have been facilitated by their accumulation of capital from opium production. The Hmong also appear to be more successful in maintaining their ethnic boundaries, even internationally, in the face of radically changed economic and social conditions, because of their non-localised clan system, while the Lua', whose identity is more closely tied to specific locations, appear to be more successful in acquiring other cultural identities. In this way, the demographic, genetic and socioeconomic patterns of population dynamics of these two 'populations', defined by traditional ethnicities in a highland Thai setting, have been very different. The material presented in this chapter illustrates the crucial significance of cultural factors in determining the responses of different groups to external forces of social and economic change.

Acknowledgements

The research discussed in this chapter was supported by a contract from the Ministry of Public Health of Thailand and grants from National Institute of General Medical Sciences, National Science Foundation (BNS 7914093, BNS 8040684), National Geographic Society, National Institute of Child Health and Human Development (RO1HD22686); Henry J. Kaiser Family Foundation, James Irvine Foundation, The California Endowment; University of California Pacific Rim Research Program, Andrew Mellon Foundation; and World AIDS Foundation. Work was carried out in collaboration with colleagues at Kasetsart, Mahidol and Chiang Mai universities in Thailand and colleagues from University of California Berkeley and University of California, San Francisco. I gratefully acknowledge the assistance throughout of Sally Lennington Kunstadter, Rasamee Thawsirichuchai, Wirachon Yangyernkun, Prasit Leepreecha Saman Pongamornkul, and the research assistants and thousands of respondents to surveys and other inquiries who contributed data to the research. Opinions expressed are those of the author.

References

Anon. (2000). 'Charting the Course. The first seven National Economic and Social Development Plans.' *Bangkok Post*, 27 February, p. 6.

Banijbatana, D. (1978). Forest policy in northern Thailand. In *Farmers in the Forest*, ed. P. Kunstadter, S. Sabhasri and E.C. Chapman, pp. 54–60. Honolulu: The University Press of Hawaii.

Boserup, E. (1965). *The Conditions of Agricultural Growth*. Chicago: Aldine Publishing Company.

Davis, K. and Blake, J. (1956). Social structure and fertility: an analytic framework. *Economic Development and Cultural Change*, **4**, 211.

Geddes, W.R. (1976). *Migrants of the Mountains: The Cultural Ecology of the Blue Miao (Hmong Njua) of Thailand*. Oxford: Oxford University Press.

Keen, F.G.B. (1978). Ecological relationships in a Hmong (Meo) economy. In *Farmers in the Forest*, ed. P. Kunstadter, S. Sabhasri and E.C. Chapman, pp. 210–221. Honolulu: The University Press of Hawaii.

Khongsanit, P. (2001). Heights of folly: encroachment and deforestation in areas of the Upper North are leaving mountains stripped of forest cover – and of the natural resources they once contained. *Bangkok Post*, 12 February.

Kunstadter, P. (1966). Residential and social organization of the Lawa of northwestern Thailand. *Southwestern Journal of Anthropology* **22**(2), 61–84. Reprinted (1969) in *Introductory Readings on Sociological Concepts, Methods and Data*, ed. M. Abrahamson. New York: Van Nostrand Reinhold Co.

Kunstadter, P. (1971). Natality, mortality and migration of upland and lowland populations in northwestern Thailand. In *Culture and Population*, ed. S. Polgar, pp. 46–60. Cambridge, MA: Shenkman Publishing Co., and Carolina Population Center.

Kunstadter, P. (1972). Demography, ecology, social structure and settlement patterns. In *The Structure of Human Populations*, ed. A. Boyce and G. Harrison. Oxford: Clarendon Press. Reprinted in *Human Ecology Series*, ed. A.P. Vayda. Andover MA: Warner Modular Publications, Inc.

Kunstadter, P. (1978). Subsistence agricultural economies of Lua' and Karen hill farmers, Mae Sariang District, northwestern Thailand. In *Farmers in the Forest*, ed. P. Kunstadter, S. Sabhasri and E.C. Chapman, pp. 74–133. Honolulu: The University Press of Hawaii.

Kunstadter, P. (1983a). Highland populations in northern Thailand. In *Highlanders of North Thailand*, ed. J. McKinnon and W. Bhruksasri, pp. 1545. Kuala Lumpur: Oxford University Press.

Kunstadter, P. (1983b). Animism, Buddhism and Christianity – religion in the life of Lua' people of Pa Pae, northwestern Thailand. In *Highlanders of North Thailand*, ed. J. McKinnon and W. Bhruksasri, pp. 135–154. Kuala Lumpur: Oxford University Press.

Kunstadter, P. (1984a). Cultural ideals, socioeconomic change, and household composition: Karen, Lua', Hmong and Thai in Northwestern Thailand. In *Households: Comparative and Historical Studies of the Domestic Group*, ed. R. McC. Netting, R.R. Wilke and E.J. Arnould, pp. 299–329. Los Angeles: University of California Press.

Kunstadter, P. (1984b). *Demographic Differentials in a Rapidly Changing Mixed Ethnic Population in Northwestern Thailand*. Tokyo: Nihon University Population Institute, NUPRI Research Paper Series 19.

Kunstadter, P. (1985). Rice in a Lua' subsistence economy. In *Food Energy in Tropical Ecosystems*, ed. D. Cattle and K. Schwerin, pp. 21–44. New York: Gordon and Breach.

Kunstadter, P. (1986). Ethnicity, ecology and mortality in northwestern Thailand. In *Anthropology and Epidemiology*, ed. C. Janes, R. Stall, S.M. Gifford, pp. 125–156. Dordrecht: D. Reidel.

Kunstadter, P. (2000). Changing patterns of economics among Hmong in northern Thailand 1960–1990. In *Turbulent Times and Enduring People. Mountain Minorities in the South-East Asia Massif*, ed. J. Michaud, pp. 167–192. Richmond, Surrey: Curzon.

Kunstadter, P. (2002). Hmong marriage patterns in relation to social change. In *The Hmong in Southeast Asia: Current Issues*, ed. G. Lee, J. Michaud, C. Culas and N. Tapp. Chiang Mai: Silkworm.

Kunstadter, P. and Kunstadter, S.L. (1990). Health transitions in Thailand. In *What We Know about Health Transition*, ed. J.C. Caldwell, S. Findley, P. Caldwell, G. Santow, W. Cosford, J. Braid and D. Broers-Freeman, Vol. 1, pp. 213–250. Canberra: Health Transition Centre, The Australian National University.

Kunstadter, P., Kunstadter, S.L., Leepreecha, P. and Podhisita, C. (1994). Infrastructural, economic and demographic change: Hmong in Thailand. *High Plains Applied Anthropologist*, **14**(2), 97–114.

Kunstadter, P., Kunstadter, S.L., Podhisita, C. and Ritnetikul, P. (1989). Hmong demography: an anthropological case study. *Proceedings of the Meeting of the International Union for the Scientific Study of Population*, Vol. 3, pp. 317–330.

Kunstadter, P., Podhisita, C., Leepreecha, P. and Kunstadter, S.L. (1991). Rapid changes in fertility among Hmong of northern Thailand. *Proceedings of the Thai National Symposium on Population Studies*, 21–22 November 1991, pp. 103–132. Salaya, Nakorn Pathom: Institute for Population Research, Mahidol University.

Leach, E.R. (1954). *Political Systems of Highland Burma.* Cambridge, MA: Harvard University Press.

Lemoine, J. (1972). *Un Village Hmong Vert du Haut Laos: Milieu Technique et Organisation Sociale.* Paris: Edition du Centre National de la Recherche Scientifique.

Nimmenahaeminda, K. (1965). An inscribed silver plate grant to the Lawa of Boh Luang. *Felicitation Volumes in Southeast Asia Studies Presented to His Highness Prince Dhanivat Kromamum Bidyalabh Bridyakorn*, **2**, 233–238. Bangkok: The Siam Society.

Ratanakorn, S. (1978). Legal aspects of land occupation and development. In *Farmers in the Forest*, ed. P. Kunstadter, S. Sabhasri and E.C. Chapman, pp. 45–53. Honolulu: The University Press of Hawaii.

Sabhasri, S. (1978). Effects of forest fallow cultivation on forest production and soil. In *Farmers in the Forest*, ed. P. Kunstadter, S. Sabhasri and E.C. Chapman, pp. 160–184. Honolulu: The University Press of Hawaii.

Sukwong, S. (1999). Legislation and community forestry 1999. *Proceedings of Symposium on Significance of Mangrove Ecosystems for Coastal People*, ed. P. Kunstadter, pp. 127–132. Mangrove Ecosystems Proceedings Number 4. Nishihara, Okinawa: International Society for Mangrove Ecosystems, Prince of Songkla University, National Research Council of Thailand and Royal Forest Department.

Yang, D. (1992). The Hmong: enduring traditions. In *Minority Cultures of Laos: Kammu, Lua' Lahu, Hmong and Mien*, ed. J. Lewis, pp. 249–326. Ranch Cordova, CA: Southeast Asia Community Resource Center, Folsom Cordova Unified School District.

Zinke, P., Sabhasri, S. and Kunstadter, P. (1978). Soil fertility aspects of the Lua' forest fallow system of shifting cultivation. In *Farmers in the Forest*, ed. P. Kunstadter, S. Sabhasri and E.C. Chapman, pp. 134–159. Honolulu: The University Press of Hawaii.

10

Ecology, homeostasis and survival in human population dynamics

ROBERT ATTENBOROUGH

Introduction

The complex processes of human population dynamics work themselves out, not in an abstract limbo but, as for other species, in particular environmental settings. The specific physical and biotic characteristics of such settings may crucially affect these dynamics and vice versa. From their local environments, and/or from any others to which personal travel and exchange systems allow access, people must acquire the material resources, such as food and water, necessary for individual and population survival. Complicating factors peculiar to *Homo sapiens* have not abolished this fundamental dependence on our environments; but they have both changed the nature and time-scale of that dependence, and made its reality harder to see.

One of these complications, with a major effect on human population–environment relations, is our capacity to reach beyond the local environment to obtain resources. In a now globalised economy, access to other environments' resources is scarcely impeded by geographical barriers or distance, though very much impeded by poverty. Even in earlier times and more straitened circumstances, flows of material necessities have commonly extended well beyond the settings local to particular communities. The potential implications for population dynamics are huge.

Other complications apply in human societies too, even within local settings. People view landscapes and resources through culturally informed spectacles. They imbue them with symbolic meanings, religious beliefs, emotional attachments, proprietorial claims, and senses of individual and collective identity. They develop and communicate sophisticated understandings of geography, seasonality and human impacts. These dimensions of the relationship with the environment are often connected with each other, and with directly instrumental actions which deploy socially organised labour and technology to obtain and process material resources. Of particular relevance here is Ester Boserup's (e.g. 1981) concept that population increase is not so much permitted by agricultural innovation, as itself a spur to social, economic and technological change. All these attributes affect how populations support themselves materially.

Here, then, are some reasons why, whilst no-one doubts ecology's relevance in principle, it has proved to be so difficult and controversial to investigate its rôle in human population dynamics. What, for example, is the carrying capacity for humans of a particular environment? And is that concept relevant to understanding how large or dense an actual human population is? So much depends, not only on the properties of the environment and the biological needs of the human organism, but also on the society's collectively organised ways of living in and from that environment, with all its material and ideational aspects, as well as with its access to more distant resources.

Few subsequent researchers have linked the biological productivity of environments as directly with population numbers as Birdsell (1953) did in a classic reconstruction of pre-European Aboriginal Australian population ecology. Annual rainfall, he argued, exhibited a strong negative correlation with 'tribal area', hence a positive one with population density, in areas not watered by coasts and major rivers. Many authors are more sceptical: Ellen (1982) reviewed some of the difficulties surrounding the study of carrying capacity in human population ecology.

My aim in this chapter is to put human population dynamics into some ecological context, and particularly to ask whether the long-term dynamics exhibit a pattern suggesting homeostasis, as ecological models would lead one to expect. Are there homeostatic feedback loops, which tend to keep population sizes or densities within a specific range relative to ecology and way of life, or which at least curb the potential for demographic change? If so, under what conditions are they most effective? I shall pursue this issue, not in a mathematically analytical way (cf. Lee 1986; Wood 1998), but by drawing on a range

of diverse local and regional examples, concentrating mainly on pre-
or non-industrial societies.

Population change and homeostasis

Whilst stable and stationary populations are useful conceptual models
in theoretical demography, real populations are never static, always
in numerical flux. At one level this flux is simply described by the
demographic balancing equation:

$$\text{Population} = \text{Previous population} + \text{Births} - \text{Deaths} \\ + \text{Immigration} - \text{Emigration}$$

This highlights the inherent improbability of additions to a population ever exactly matching its losses. Furthermore, the factors affecting fertility, mortality and migration generally have little in common,
and much day-to-day research into them is conducted separately by
different specialists. But are there, below the surface, connections between them? Granted that populations fluctuate, do they do so as
erratically and extremely as they would if the components of growth
were fully autonomous? Or does this change, viewed in the long term,
have the character of fluctuation around a 'target' level or within a
viable range? Granted also that, even in non-human species, population dynamics are too complex to be fully understood through the
deterministic application of a single principle such as homeostasis
(Anderson *et al.* 1979; Cappuccino and Price 1995) and that, furthermore, we have the peculiar features of the human species to take into
account, is there an element of ecological and/or economic homeostasis at work?

Concepts such as homeostasis, regulation and adaptation have a
more secure and respected place in a biologist's vocabulary than in a
social scientist's (Smith 1993; Ulijaszek 1997). In the mainstream of
sociocultural anthropology, for example, the concept of adaptation
lost much of its standing with the demise first of nineteenth-century
evolutionism and then of early twentieth-century functionalism
(Morphy 1993). Furthermore, even for a population biologist, these
concepts are more problematic than they are for, say, a physiologist.

That an organism's cellular and organ systems regularly interact adaptively and homeostatically is uncontroversial (Harrison 1993), but when a biological population is under consideration – just as when a society or its culture is – such immediate, rigorous self-regulation cannot be taken for granted. The perpetuation of populations and societies is not self-evidently dependent on maintaining absolute population numbers – or population density, or any other demographic variable – within limits as tight as those of physiological homeostasis. Empirically, the global human population and many national ones have, for part or all of the modern era, grown in seemingly runaway fashion (McEvedy and Jones 1978; Livi-Bacci 1997), and modern debates about human population are peppered with anxieties about its potentially disastrous proneness to change. So need we even ask about the power of human population dynamics to resist pressures for decline or growth?

Perhaps not, if we only consider industrial and post-industrial mass societies. But these are the societies where the components of population change are most likely to become separated. Do we get a different answer when we consider the long run of human history and the full range of sociocultural diversity? Amongst writers since Malthus (1970 [1798]) who have thought so, Carr-Saunders (1922) developed a complex theory of human population regulation, evidenced from the anthropology of his day. Cohen *et al.* (1980) drew together many threads from subsequent research, but over the last 15 years or so there has been considerable renewal of interest (e.g. Coleman 1986; Lee 1986; Wood 1998; Wilson and Airey 1999).

Human demography and animal population ecology

In demography, models of a population's workings and projections of its future are conventionally built on data and assumptions about the core demographic variables themselves – fertility, mortality and migration, and sometimes about economic conditions – rarely on ecology explicitly. A somewhat double-edged example was provided by Coale (1974: 25) when he projected the contemporary global population growth rate indefinitely, to the outcome that 'in less than 6,000 years the mass of humanity would form a sphere expanding at the speed of light'. His main point was to show, dramatically, that the contemporary growth rate could not continue indefinitely. His intentional *reductio ad absurdum* both illustrated standard demographic projection methods and showed the lack of reality in those methods in the long run, when based on demographic variables alone.

Animal population dynamics are treated rather differently. Natural historians have long been impressed by the persistence and relative numerical constancy of many natural populations (as well as the wild fluctuations of some), and have posed questions about population survival and regulation. There is a long-standing, if often tacit, assumption that population phenomena are to be understood in an ecological framework (e.g. Andrewartha and Birch 1954; Cappuccino and Price 1995). The focus is on how the abundance and distribution of one species interact with those of others, and with other relevant environmental resources and hazards.

Many ecological models incorporate negative feedback properties, whereby the abundance of a particular species, instead of moving indefinitely in a given direction, tends towards or oscillates around an equilibrium. This is because there are density-dependent (as well as density-independent) impacts on animal numbers. An increase in prey abundance permits an increase in predator or parasite abundance, which in turn places a brake on prey abundance, and so forth. Density-dependent negative feedbacks thus lead to homeostasis. The relative importance of density-dependent factors versus density-independent ones, and the applicability of equilibrium models, are debated on both empirical and theoretical grounds, but both have strong support and they are not mutually exclusive (Cappuccino and Price 1995: e.g. p. 12). Their relative importance also depends on the time-scale investigated.

Expected ranges for demographic variables

Demographic variables such as fertility, mortality and growth have finite numerical ranges. On a non-equilibrium model, we might predict that observed variation over time in a demographic variable will take the form of a 'random walk' through the feasible range for that variable, eventually leading in some cases to population extinction. On an equilibrium model, we might predict instead that observed variation is usually confined to a limited part of the feasible range, with lower risk of extinction. A first, non-mathematical, step towards discriminating between these models would be to characterise these feasible ranges. We need to specify numbers, since what counts as 'high' or 'low' in one society or at one period of history may not in other situations, and what is empirically average may be extreme from a theoretical viewpoint.

If pregnancy takes nine months and ovulation may return very quickly, then, naïvely speaking, a woman who lives and is reproductively active from 15 to 45 years might be able to produce 30 to 40 children. Although such levels are occasionally reached and exceeded by individual women, especially through repeated multiple births, they are far beyond levels ever recorded as population averages, such as 8.9 (North American Hutterites, 1950: Bongaarts 1975). Whilst even 8.9 is extreme, many recorded *population completed family sizes* (CFS) and *total fertility rates* (TFR) are well above the levels familiar to most residents of twenty-first-century developed countries. In Africa many current national TFRs range up to 6.8 to 7.2 (Haub and Cornelius 2000). Nearly all developed countries nowadays have TFRs below the replacement level of 2.1, whether marginally (e.g. Iceland, 2.05) or substantially (e.g. Latvia, 1.09) (Monnier 1999). CFSs for locally defined populations range at least from 3.7 to 6.1 for specific historical European populations, from 6.1 to 9.5 for religious isolates in North America, 5.3 to 6.7 for various non-European populations, and from 2.6 to 8.4 for hunting and gathering populations (Gage *et al.* 1989; Pennington 2001).

Maximum fertility levels are generally found in societies practising *natural fertility* régimes. Natural fertility does not correspond to a single TFR but to a range, 3.5 to 9.5 (mean 6.1: Wood 1990). Varying proximate determinants of fertility contribute to this variation natural fertility encompasses non-deliberate determinants such as age at marriage, proportion not marrying, and the biological impacts of breastfeeding, and maybe undernutrition, on amenorrhoea and birth-spacing (Wood 1990). Except in populations with very high prevalence of sexually transmitted disease, it appears hard to reduce TFR below about 4 without either large-scale contraception or abortion, or very low coital frequency. The highest TFRs are those of natural fertility groups living under affluent conditions with little resort to breastfeeding.

Mortality levels familiar to most residents of twenty-first-century developed countries are likewise unrepresentative of the full range of demographic régimes. Whereas currently life expectancies in these countries are over 70, or even 80, years (Haub and Cornelius 2000), the people in contemporary developing countries and 'anthropological' societies generally have life expectancies in the range 30–65 years, as did the now-developed countries before the nineteenth century (Gage *et al.* 1989). In pre-industrial settings, 40 years would have been a high life expectancy: figures in or even below the 30s were more usual. Possibly the lowest life expectancies to emerge from documentary evidence are those of 17 years for plantation slaves in Trinidad, 1813–16 (John 1988). The Yanomama hunter-horticulturalists of Venezuela have life expectancies variously estimated at 20–21 and even 15–17 years (Gage *et al.* 1989). Archaeologically derived estimates of ancient life expectancies are especially fraught with methodological difficulties; but Hassan's (1981) overview suggests that they were often in the low 30s from the Neolithic times with only marginal gains up to the eighteenth century.

Annual rates of natural increase in present-day national populations extend from the negative range (e.g. −0.6% in Russia) to a few countries where it is over 3% (e.g. 3.3% in Chad), with a world average currently at 1.4% (Haub and Cornelius 2000). Thus, annual growth at the hardly impressive-sounding rate of 2% is high (doubling time 34.7 years), and 4% (doubling time 17.3 years) would be extraordinary. Populations in 'frontier' situations are notable for rapid growth, for example European farming settlers in Australia, 1861–1900, whose natural increase was often well over 2% and on occasion approached 3% (Borrie 1994: 130–131); with migration, it was even higher.

The current global growth rate and its momentum, as well as many national rates, are sufficiently far from zero to imply major population change within decades. But any non-zero growth rate, even one that would be difficult to distinguish from zero, still implies major population change when projected over a sufficiently long time period. The arithmetic involved takes no account of either stochastic fluctuation or homeostatic oscillation. Writing about Aboriginal Australia, Gray (1985) points out that annual rates of growth which might be classified as 'near-stationary', e.g. up to 0.2%, can still produce a several-fold increase over a thousand years or 50-fold over 2,000 years. A rate of 0.1% could have produced, even from a low base of 10,000 people in Australia 8,000 years ago, an Aboriginal population of 30 million in AD 1788, far higher than any serious estimate. That Australia was populated for over 40,000 years and had in 1788 an Aboriginal population for which one million is a high estimate implies very low net growth indeed. Gray concludes (p. 26): 'we should expect to find very considerable changes over time in Aboriginal population sizes (both local and as a whole). Such changes are relatively insignificant and do not contradict the hypothesis of long-term stationarity.' Hassan (1981) arrived at an average world estimate of 0.1% for Neolithic population growth. An average annual growth rate of 0.015% over the last 100,000 years would be sufficient to produce the present-day population of 6 billion (Pennington 2001).

What level of fertility is required to balance a given mortality régime? In terms of the balancing equation, crude birth and death rates must simply be equal in a closed population; but these measures are potentially misleading. Based on some reasonable assumptions about the age-patterning of mortality and fertility (Coale–Demeny West model life table, typical fertility pattern), the expected mathematical relationship between life expectancy and TFR in a stationary population, derived from Coale *et al.* (1983), can be summarised as:

Life expectancy at birth (years)	20	25	30	35	40
Total fertility rate (children)	6.6	5.3	4.5	3.9	3.4

A life expectancy of 30 years would seem drastically low to a twenty-first century reader in any country (Haub and Cornelius 2000); but a TFR in the region of 4.5 is sufficient to counter-balance it – not in any sense pushing fertility to its limits.

Closed ecosystems and small remote places

Mountain valleys, remote islands and clusters of rainforest hamlets are attractive microcosms in which to envisage the working out of population–environment dynamics. People and resources may seldom move in or out of the local environment; feedback loops may be short, local and fast-acting; traditional practices may promote homeostasis; and demographic and ecological variables may be on a limited scale, amenable to measurement. How does this work out?

In some such cases any homeostasis simply fails, as it did for the now extinct Norse Greenlanders (McGovern 1980). The Polynesian Henderson Islanders, too, disappeared, having apparently exterminated six bird species on which they had depended; and the Easter Islanders also suffered acute depopulation following unsustainable resource use (Diamond 1991).

The outcomes of isolation are not always so melancholy; but many things can frustrate a researcher's hope of seeing how successful homeostasis is achieved. Bathurst and Melville Islands in the Arafura Sea were evidently inhabited before the rise in sea level which by 8,000–10,000 years ago had separated them completely from mainland Australia. The Tiwi Aboriginal people of these islands lived there for millennia, apparently without experiencing critical resource stress or population pressure. We have only one reliable census, Hart's in 1929 (Peterson and Taylor 1998), to represent autonomous Tiwi population ecology before outside contacts transformed it; even though it is one of the best hunter-gatherer censuses available, it is a slim basis for inference in the present context. The Zangskar Valley in the western Himalayas is another remote and demographically successful population, but its isolation is relative and its demographic–environment system far from exhaustively studied (Crook and Osmaston 1994). Another rigorous montane environment, the Nuñoa district in the Peruvian Andes, less rugged but very high, has been more comprehensively studied ecologically, but it is still further from being closed to human and resource movement. High-energy foods were imported into Nuñoa in exchange for low-energy animal products and some 76% of the food energy consumed by a 'typical'

Nuñoan family was imported into the local ecosystem (Thomas 1976, 1997). Thus historical circumstances and data limitations often pose significant blocks to our view of populations that would, ironically, be of particular interest.

In some cases, however, more information may be available. The 1,820-strong Gidra of lowland southern Papua New Guinea, between the Fly River and the Torres Strait, are not physically cut off from neighbours; but even in 1980 there was, as often in Melanesia, high linguistic endogamy. Ohtsuka (1986: 14) considered that they could be treated as 'an entity of survival', i.e. essentially closed, at least until recently, and he based his research on oral genealogical knowledge. He concluded that until 1945 the Gidra long-term mean annual growth rate had been 0.2% (doubling time 347 years), which is within Gray's (1985) category of 'near-stationary' levels of population growth. Calculation backwards from 1945 to the foundation of the population, possibly 300 years earlier, yielded an estimate of 713 for the size of the founding population. The historical Gidra thus appear to exemplify low growth and relative homeostasis. Gidra women marry soon after menarche, usually remarry after widowhood or divorce during their reproductive careers, and seldom if at all practise birth control or infanticide, generally desiring as many children as possible. Ohtsuka discusses factors possibly involved in the low mean fertility that none the less results, of which the most significant appears to be high infecundity not apparently due to sexually transmitted disease, with malaria and malnutrition as background factors.

Dye and Komori (1992), using a method based on the frequency distributions of radiocarbon dates, generated for the Hawaii archipelago a curve which estimates the 'pre-censal' growth of the Hawaiian population from its foundation before AD 400 to its first credible census in 1831–2. That census provides absolute numbers by which to calibrate the curve, yielding population estimates growing from 1,295 in AD 332 to 163,293 in AD 1441, then declining and partially recovering to 132,338 in 1778, when Cook's fleet arrived. This last population figure is not greatly different from the 1832 figure of 130,313. Thus, if the method is accepted, three phases emerge: (a) slow growth to the twelfth century, (b) quite rapid growth to the fifteenth century, and (c) fluctuation without net growth to the early nineteenth century. Despite the picture of growth that this presents, the annual growth rates implied are not extreme by modern standards: 0.32% when calculated from AD 400 to 1778, 0.93% in the period of fastest growth 1200–1340 (doubling times of 217 and 75 years respectively).

In an ecological hypothesis to account for this growth, the founding population is envisaged as arriving in a vacant, productive and healthy environment, populating it, and eventually encountering density-dependent brakes as the new niche 'filled up'. Infanticide, celibacy, abortion, *coitus interruptus*, sea-voyaging and war are considered as brakes, whilst a need to sustain a larger population also led to agricultural intensification in the sixteenth and early seventeenth centuries. But a density-independent hypothesis has also been argued: that intensified production was more the product of political competition amongst status rivals seeking to produce surplus food. Either way, after a millennium of moderate growth, the Hawaiian population appeared to have reached a plateau lasting several centuries. Neither faunal extinctions nor social nor agricultural changes in this latter period seem to have been sufficient to destabilise a good candidate for population homeostasis.

Despite a lack of time depth in the available data, then, some remote communities are clearly less cut off than some of us have romantically imagined. None the less, there are hints here that non-industrial populations have sometimes – not always – settled isolated and difficult environments very successfully, and have developed ecologically sustainable ways of life there, with long-term demographic homeostasis or very slow population growth.

Historical demography as population ecology

The cosmopolitan mass societies of Europe and Asia have long since ceased to be locally self-sufficient as regards the organisation of populations or the flow of resources; and their complex population dynamics are correspondingly hard to envisage in ecological terms. However, in some cases, these societies hold a priceless asset for the understanding of population ecology: a rich historical depth in the available data, which is seldom or never matched in more self-contained settings (Swedlund 1978). The historical data that make this possible are demographic and sometimes economic, but very seldom directly ecological. None the less, there is a key ecological issue at stake. Can it be shown, either at a continental and secular scale that population growth is dampened relative to its potential, or at a local and decadal scale that people regulate their family formation relative to their aspirations and circumstances, rather than leaving it solely to biology? In cases where the answer is 'yes', we may arguably be seeing at least the shadow of unrecorded background ecological pressures and opportunities passing across the lives of those people whose only written records are in the historical demography. In this section I draw particularly on Wilson and Airey's (1999) synthesis of the evidence for long-run relative population homeostasis in Europe and East Asia. In the following section I briefly extend the overview to the less well known case of Java in the nineteenth century.

Currently the most thoroughly researched historical demographic records are those for Europe (see also Smith, Chapter 6), followed by those for East Asia, especially China. Long-run growth rates through the Christian Era are low in both, up to AD 1700 (Wilson and Airey 1999). China's overall average annual growth rate from AD 2 to 1500 is approximately 0.03% (doubling time 2310 years). Viewed in shorter time periods, both regions manifest fluctuation concealed by the average, before 1700 mainly ranging from −0.2–0.3% to +0.2–0.3%, though with more centuries of growth than of decline. This variation is attributed to factors such as climate and varying political stability. Episodes of excess mortality, notably the Black Death pandemic in the fourteenth century, do affect the growth rates, but even this constitutes a downward inflection rather than a massive check. Long-term growth is modest and fairly steady, implying that fertility control and migration opportunities played important rôles in influencing growth, catastrophic mortality less so (Livi-Bacci 1997; Wilson and Airey 1999). Wilson and Airey, therefore, favour the view that, although growth over these periods was not zero, societies in both regions had developed regulatory mechanisms, which kept them close to that level, and kept other demographic variables within similar ranges – TFRs of four to six children, which is not far from balancing life expectancies at birth of 25 to 40 years.

The nature of these regulatory mechanisms was strikingly different, even though the resulting demographic régimes were similar. The 'European marriage pattern' (Hajnal 1982) involves: relatively late female marriage; relatively high percentage of non-marriage; and, within marriage, relatively unconstrained reproduction, close to natural fertility, although the duration of breastfeeding was important for fertility outcomes.

In China, during broadly the same period, marriage was at generally younger ages and near-universal, so that marriage itself was not a brake on fertility (Zhao 1997). However, non-technological forms of fertility control operated within marriage, as well again as prolonged breastfeeding on the part sometimes of malnourished mothers. The nature of such fertility control is uncertain, but demographic evidence shows longer birth-spacing than in Europe, younger age at last birth, and differential marital fertility according to number and sex of children already born. This implies that 'stopping rules' were not confined to the literate. Even as late as China's 1982 One-Per-Thousand-Population Fertility Survey of women born between 1914 and 1930, women who already had both sons and daughters, and especially those whose sons outnumbered daughters, were least likely to proceed to further children, and had the longest inter-birth intervals and the lowest ages at last birth (Zhao 1997). Controls may have operated through the frequency or timing of coitus, including a gap between marriage and the onset of childbearing, and 'terminal abstinence'. Other behavioural factors may have affected numbers of surviving children per family, including differential child care according to sex, of which there is again indirect evidence, at times including infanticide affecting both sexes to varying degrees. Lee and Wang (1999) summarise the salient features of 'endogenous restraint' on population growth under four main headings: predominantly female infanticide and other excess female mortality; male celibacy; adoption and fictive kinship; and marital restraint. There was also exogenous stress due to climatic, economic or epidemic causes. Not only was Chinese traditional culture apparently not as pro-natalist as sometimes supposed, but people were exerting choices in childbearing behaviour in response to socioeconomic conditions (Zhao 1997).

Neither Europe nor East Asia was in general so hard-pressed by mortality that population homeostasis required the maximum fertility of which populations were biologically capable (see above). Malthus's antinomy, of Europe being dominated by the 'preventive check' of nuptiality, whilst other regions, including China, were dominated by the 'positive check' of mortality, was wide of the mark (Lee and Wang 1999; Smith, Chapter 6). Japan resembles China more than Europe, though the resemblance is not exact.

Migration between environments can itself be a response to pressure. Amongst other striking aspects of both European and Chinese demography at this period are rural–urban migration (see Clarke, Chapter 3) and rural–urban mortality differentials. In Britain much of the natural increase from the rural areas migrated to cities, without whom, since urban life expectancy was so poor, the cities could scarcely have persisted, let alone grown (Wrigley 1988). This may have applied both elsewhere in Europe and in China.

Wilson and Airey (1999) are impressed by the extent to which there is potential for homeostatic processes to supply models for future research as more detailed bodies of historical demographic data become available. They argue that, especially when surviving as well as ever-born family sizes are considered, low net fertility and low population growth have long been part of the sociocultural and economic systems of both regions. For this reason, culturally conservative attachment to high fertility cannot be a long-standing or insuperable block to the functioning of societies with either a European or a Chinese type of demographic régime.

The case of historical Java

Like Europe and East Asia, island south-east Asia also has substantial historical data on pre-industrial population dynamics. Here I briefly review the complex case of Java, with a view, as in the previous section, to considering whether hints of population- and family-level homeostasis suggest the shadowy presence of background ecological processes not recorded in detail. I aim not only to test some ideas aired in previous sections, but also to illustrate complicating factors of types not raised so far.

The population dynamics of Java have been under intensive discussion ever since Geertz's (1963) controversial interpretation of 'agricultural involution' and the capacity, up to very high levels, of wet-rice cultivation ecology to reward increased labour input with increased yields (see also Weiner 1972; Alexander 1984). Reid (1988) paints a background picture of south-east Asia's population as sparse in AD 1600 (half Europe's density, one-sixth of China's) and experiencing growth slower than Europe's or China's through the seventeenth and eighteenth centuries. For Java, he estimates a denser population of 4 million in 1600, but also growing slowly (0.11%, doubling time 630 years) to perhaps 6 million in 1815 (Reid 1987). Boomgaard (1989) estimates nineteenth-century growth at 1.25% (55.4 years) before 1850 and 1.6% afterwards (43.3 years). From 19.5 million in 1880 (Boomgaard 1989), the population grew to 29 million in 1900, 35 million in 1920, and 41.7 million in 1930, with growth rates between 1% and 1.5% (69.3 and 46.2 years) between 1880 and 1942, when the colonial era ended (Hugo et al. 1987; Gooszen 2000). The picture is one of accelerating growth until the 1970s when modern birth control was taken up on a large scale. Today the population is over 120 million.

The historical and economic background to pre-colonial and colonial growth, of changing population geography and subsistence practices, warfare amongst Javanese polities, and the rising economic, military and governmental power of Dutch commerce and colonialism, including the 'forced cultivation system', is detailed by a number of historians (Ricklefs 1986, 1993; Reid 1993; Elson 1994). The gathering pace of population growth through the nineteenth century was unevenly patterned geographically, though generally population density and agricultural intensity increased hand in hand. Wet-rice cultivation expanded into areas not previously cultivated in that mode (see also Kunstadter, Chapter 9), and population expanded up the hills in some areas, whilst also abandoning less productive areas. From core areas in central Java – where population growth had accelerated early and consequently the shortage of arable land came to be felt early as well – the frontier expanded into east Java from the 1850s and especially from 1880. By 1930, all suitable arable land in central and east Java was brought into production. Development in west Java also proceeded, gradually and in pockets, to fill all suitable areas.

The Geertzian argument that population growth was fuelled by peasants' need for large families has proved controversial. According to this view the motive was to supply labour for large-scale commerce as well as food cropping, and the mechanism, in some versions, was a decline in breastfeeding related to women's agricultural work, and hence also in post-partum sexual abstinence (Alexander 1984). That rising fertility was the main component of growth at all is contested by scholars who ascribe it a slighter rôle than falling mortality (e.g. Elson 1994), or even find fertility (crude birth rate) to be virtually unchanging over the period (Boomgaard 1989). Where falling mortality is the main explanation, that is put down (Boomgaard 1989; Elson 1994) to a combination of:

- *Pax Neerlandica*, a favourite explanation of contemporary colonial officals but whose benevolence is contested (Nitisastro 1970);
- better communication, increasing food supply and availability of income to purchase it, hence improving nutrition;
- de-urbanisation of the population; and
- the intervention of the colonial medical authorities, which was only clearly effective against one of the five main fatal infections, smallpox, for which there was mass vaccination from 1816.

Around 1880–90, however, whilst economic and population growth continued, scope for expansion was running out, and increasingly signs emerged of impoverishment amongst certain groups – landlessness, hardship amongst wet-rice leaseholders and even, in places, famine compounded by outbreaks of infectious disease (Elson 1984; Gooszen 2000). As the laterally expanding frontier reached its limits, increasing agricultural production and population growth had to be supported principally by intensification and economic change, with considerable success for some decades (van der Eng 2000).

Marriage was near-universal. The ages given for first marriages are young, usually between 14 and 18 years for females, although there was variation linked to access to land (Boomgaard 1989). Divorce, remarriage and adoption were common. Boomgaard (1989) found a partial link between age at first marriage and birth rate, thus a partial echo of the European pattern. He also proposed, on the basis of regional comparisons, that Javanese couples had a 'target' family composition in mind, which was sensitive to both mortality régimes and economic (including non-agricultural) opportunities. Thus he inferred that they were able and willing to practise birth control, and traditional methods, of uncertain efficacy but frequently used, and use of various abortifacients, were indeed reported. *Coitus interruptus* is sanctioned by Islam and widely known. Infanticide did not occur. Other factors with a potential to affect fertility included:

- sexual abstinence, sanctioned in various circumstances not linked to deliberate fertility control and not generally prolonged;
- breastfeeding normally for two years or so; and
- high divorce rate.

Boomgaard concluded further that, since fertility was under their control, continuing higher fertility to mid-century despite falling mortality was a conscious choice by Javanese women or couples; and that subsequent small reductions in fertility may have resulted from both marriage pattern changes and more stringent birth control. Overall, Boomgaard concluded that, rather than population growth driving economic development in keeping with a Boserupian hypothesis, both were the outcome of

> the fact that the colonial state was able to carry out its ambitious programme of increased production for the world market, through a combination of compulsory cultivation services, an improved infrastructure, and an effective vaccination campaign
>
> (Boomgaard 1989: 203).

Some corroboration comes from a small-scale study of the 1883 famine records for Indramayu and Kadanghaur, west Java (M.R. Fernando, personal communication). It appeared that marriage there was sometimes delayed (for men only, however), and more significantly that there were delays between marriage and the start of childbearing, as well as an early cessation of childbearing, relative to biological potential. Both of these imply, perhaps, a larger rôle than suggested by Boomgaard, in some regions at least, for sexual abstinence. There were also indications of a preference for male children in determining when childbearing ceased. Hull (2001) reviews Boomgaard's conclusions for Java as a whole, also raising amongst other issues the possible rôle of abstinence as an intentional fertility control measure.

Many of these patterns remain to be confirmed by more comprehensive studies, but there are hints here of an element of population regulation in the midst of rapid growth; of couples making their own choices rather than being forced to the limits of their reproductive potential; and of differences in the process compared to both Europe and China. In the end, however, under pressures initiated at least partly by colonialism, growth outstripped homeostasis until Java's recent fertility transition.

Conclusion

In this chapter, with its emphasis on the ecological perspective, I may have seemed to sideline a number of factors that are very important for the spatial distribution and temporal dynamics of populations in their environments, in particular the sociocultural factors emphasised by Kunstadter (Chapter 9). Forge (1972), for example, collated data on New Guinea settlement size, which tends to range from 70 to 300 individuals. He accepted that ecological factors in difficult terrain play a part in explaining the small social units and that warfare plays a part in the exceptional cases where the settlements are larger, but he believed that these were not the only factors. He based his arguments on the requirements of the type of egalitarian society found in Melanesia, and he proposed that where the social unit rises above 350 to 400, or 80 to 90 adult males, egalitarian society becomes unmanageable, because of the number of face-to-face relationships that need to be maintained. So, as this case illustrates, social factors may often be crucial in affecting both settlement patterns and subsistence ecology, and should always be considered in studies of human population dynamics. Yet, in another contrasting example, Netting

(1993) described how the Kofyar people of the Jos Plateau, Nigeria, abandoned familiar intensive agricultural techniques for slash-and-burn agriculture on migration to sparsely settled frontier lands, reverting to intensive techniques within 30 years as population density increased, with concomitant changes in household composition. What is shown, therefore, is the interaction of factors.

The main emphasis in this chapter, however, has been on long-term population numerical homeostasis amidst short- and medium-term flux; on the ecological factors on which homeostasis is usually presumed to depend and the economic ones which mediate them; and on the resultant prospects of long-term population survival. Following Lee (1986, 1994), Wood (1998) and especially Wilson and Airey (1999), I have selectively explored points about, and some illustrations of, the homeostatic elements in human population dynamics, with special reference to non- or pre-industrial populations. It seems timely to follow their call for renewed attention to the possible interconnections amongst components of the demographic régime – in fact, the ecological–demographic system as a whole. This system may incorporate a degree of homeostasis, however loose and contingent, which we, in the twenty-first century, would do well to keep within our sights. Differences of opinion on the relative importance of density-dependent and density-independent factors may in part reflect different emphases on longer or shorter time-scales, since density-dependence theoretically becomes more important as the time-scale is increased. We have, I suggest, seen evidence of long-term density-dependence in the examples reviewed here and, even if the long term means millennia, this is food for thought.

Population growth, in most historical circumstances, has been at low or moderate levels, even in the absence of catastrophic mortality. This reflects, ultimately and through the complex mediation of sociocultural and economic systems, the environmental constraints and opportunities with which, in our different ways, we alongside other mammals must live. It may also reflect the readiness of both human biological systems and human cultural systems generally to respond adaptively to changing circumstances, including humanly generated ones, as well as their compatibility with a wide range of specific environments, niches and economic modes, and with demographic régimes of low growth and small effective family size. For perhaps longer than we conventionally recognise, human beings have not merely been buffeted by unavoidable mortality and come to accept the desirability or inevitability of frequent pregnancy. In addition, there are, in Livi-Bacci's (2000: 1) phrase, both 'factors of constraint and factors of choice'. Cultural family systems have both intended (e.g. abstinence and contraception) and unintended (e.g. care of the elderly by unmarried daughters) consequences at the level of overall population maintenance. Those family systems in turn, though not environmentally determined, may often be sensitive to ecological factors relevant to their wider communities, societies and populations.

References

Alexander, P. (1984). Women, labour and fertility: population growth in nineteenth century Java. *Mankind*, **14**, 361–372.

Anderson, R.M., Turner, B.D. and Taylor, L.R. (ed.) (1979). *Population Dynamics: 20th Symposium of the British Ecological Society, London 1978*. Oxford: Blackwell Scientific.

Andrewartha, H.G. and Birch, L.C. (1954). *The Distribution and Abundance of Animals*. Chicago: University of Chicago Press.

Birdsell, J.B. (1953). Some environmental and cultural factors influencing the structuring of Australian Aboriginal populations. *American Naturalist*, **87**, 171–207.

Bongaarts, J. (1975). Why high birth rates are so low. *Population and Development Review*, **1**, 289–296.

Boomgaard, P. (1989) *Children of the Colonial State: Population Growth and Economic Development in Java, 1795–1880*. Centre for Asian Studies Amsterdam Monographs, **1**. Amsterdam: Free University Press.

Borrie, W.D. (1994). *The European Peopling of Australasia: a Demographic Enquiry, 1788–1988*. Canberra: Demography Program, Australian National University.

Boserup, E. (1981). *Population and Technology*. Oxford: Basil Blackwell.

Cappuccino, N. and Price, P.W. (ed.) (1995) *Population Dynamics: New Approaches and Synthesis*. London: Academic Press.

Carr-Saunders, A.M. (1922). *The Population Problem: a Study in Human Evolution*. Oxford: Clarendon Press.

Coale, A.J. (1974). The history of the human population. In *The Human Population* (originally appeared as September 1974 issue of *Scientific American*). San Francisco: W.H. Freeman.

Coale, A.J., Demeny, P. and Vaughan, B. (1983). *Regional Model Life Tables and Stable Populations* (2nd edn). Studies in Population. New York: Academic Press.

Cohen, M.N., Malpass, R.S. and Klein, H.G. (ed.) (1980). *Biosocial Mechanisms of Population Regulation*. New Haven, CT: Yale University Press.

Coleman, D. (1986). Population regulation: a long-range view. In *The State of Population Theory*, ed. D. Coleman and R. Schofield. Oxford: Basil Blackwell.

Crook, J.H. and Osmaston, H.A. (ed.) (1994). *Himalayan Buddhist Villages*. Bristol: University of Bristol Press.

Diamond, J. (1991). *The Rise and Fall of the Third Chimpanzee*. London: Radius.

Dye, T. and Komori, E. (1992) A pre-censal population history of Hawai'i. *New Zealand Journal of Archaeology*, **14**, 113–128.

Ellen, R. (1982). *Environment, Subsistence and System*. Themes in the Social Sciences. Cambridge: Cambridge University Press.

Elson, R.E. (1984). *Javanese Peasants and the Colonial Sugar Industry: Impact and Change in an East Java Residency, 1830–1940*. Asian Studies Association of Australia, Southeast Asia Publications Series. Singapore: Oxford University Press.

Elson, R.E. (1994). *Village Java under the Cultivation System, 1830–1870*. Sydney: Allen and Unwin.

Forge, A. (1972). Normative factors in the settlement size of Neolithic cultivators (New Guinea). In *Man, Settlement and Urbanism*, ed. P.J. Ucko, R. Tringham and G.W. Dimbleby. London: Duckworth.

Gage, T.B., McCullough, J.M., Weitz, C.A., Dutt, J.S. and Abelson, A. (1989). Demographic studies and human population biology. In *Human Population Biology: a Transdisciplinary Science*, ed. M.A. Little and J.D. Haas. New York: Oxford University Press.

Geertz, C. (1963). *Agricultural Involution: the Processes of Ecological Change in Indonesia*. Berkeley: University of California Press.

Gooszen, H (2000). *A Demographic History of the Indonesian Archipelago 1880–1942*. Singapore: Institute of Southeast Asian Studies.

Gray, A. (1985). Limits for demographic parameters of Aboriginal populations in the past. *Australian Aboriginal Studies*, **3**(1), 22–27.

Hajnal, J. (1982). Two kinds of preindustrial household formation system. *Population and Development Review*, **8**, 449–494.

Harrison, G.A. (1993). Physiological adaptation. In *Human Adaptation*, ed. G.A. Harrison. Biosocial Society Series, **6**. Oxford: Oxford University Press.

Hassan, F.A. (1981). *Demographic Archaeology*. Studies in Archaeology. New York: Academic Press.

Haub, C. and Cornelius, D. (2000). *2000 World Population Data Sheet*. Washington, DC: Population Reference Bureau.

Hugo, G.J., Hull, T.H., Hull, V.J. and Jones, G.W. (1987). *The Demographic Dimension in Indonesian Development*. Singapore: Oxford University Press.

Hull, T.H. (2001). Indonesian fertility behaviour before the transition: searching for hints in the historical record. In *Asian Population History*, ed. Ts'ui-jung Liu. Singapore: Oxford University Press.

John, A. (1988). *The Plantation Slaves of Trinidad, 1783–1816: a Mathematical and Demographic Enquiry.* Cambridge: Cambridge University Press.

Lee, J.Z. and Wang, F. (1999). *One Quarter of Humanity: Malthusian Mythology and Chinese Realities, 1700–2000.* Cambridge, MA: Harvard University Press.

Lee, R.D. (1986). Malthus and Boserup: a dynamic synthesis. In *The State of Population Theory: Forward from Malthus*, ed. D. Coleman and R. Schofield. Oxford: Basil Blackwell.

Lee, R.D. (1994). Human fertility and population equilibrium. In *Human Reproductive Ecology: Interactions of Environment, Fertility and Behavior*, ed. K.L. Campbell and J.W. Wood. **709**. New York: New York Academy of Sciences.

Livi-Bacci, M. (1997). *A Concise History of World Population*, 2nd edn. Translated by C. Ipsen. Oxford: Blackwell.

Livi-Bacci, M. (2000). *The Population of Europe.* The Making of Europe, Oxford: Blackwell.

McEvedy, C. and Jones, R. (1978). *Atlas of World Population History.* London: Allen Lane.

McGovern, T.H. (1980). Cows, harp seals and church bells: adaptation and extinction in Norse Greenland. *Human Ecology*, **8**, 245–275.

Malthus, T.R. (1970 [1798]). *An Essay on the Principle of Population.* Harmondsworth, UK: Penguin.

Monnier, A. (1999). La conjoncture démographique: l'Europe et les pays développés d'outre-mer. *Population*, **54**, 745–774.

Morphy, H. (1993). Cultural adaptation. In *Human Adaptation*, ed. G.A. Harrison. Biosocial Society Series, **6**. Oxford: Oxford University Press.

Netting, R.M. (1993) *Smallholders, Householders: Farm Families and the Ecology of Intensive, Sustainable Agriculture.* Stanford, CA: Stanford University Press.

Nitisastro, Widjojo (1970). *Population Trends in Indonesia.* Ithaca, NY: Cornell University Press.

Ohtsuka, R. (1986). Low rate of population increase of the Gidra Papuans in the past: a genealogical–demographic analysis. *American Journal of Physical Anthropology*, **71**, 13–23.

Pennington, R. (2001). Hunter-gatherer demography. In *Hunter-gatherers: an Interdisciplinary Perspective*, ed. C. Panter-Brick, R.H. Layton and P. Rowley-Conwy. Biosocial Society Symposium Series, **13**. Cambridge: Cambridge University Press.

Peterson, N. and Taylor, J. (1998). Demographic transition in a hunter-gatherer population: the Tiwi case, 1929–1996. *Australian Aboriginal Studies*, **16**(1), 11–27.

Reid, A. (1987). Low population growth and its causes in pre-colonial southeast Asia. In *Death and Disease in Southeast Asia: Explorations in Social, Medical and Demographic History*, ed. N.G. Owen. Singapore: Oxford University Press.

Reid, A. (1988). *Southeast Asia in the Age of Commerce 1450–1680: 1, the Lands below the Winds.* New Haven, CT: Yale University Press.

Reid, A. (1993). *Southeast Asia in the Age of Commerce 1450–1680: 2, Expansion and Crisis.* New Haven, CT: Yale University Press.

Ricklefs, M.C. (1986). Some statistical evidence on Javanese social, economic and demographic history in the later seventeenth and eighteenth centuries. *Modern Asian Studies*, **20**, 1–32.

Ricklefs, M.C. (1993). *A History of Modern Indonesia since c. 1300*, 2nd edn. London: Macmillan.

Smith, M.T. (1993). Genetic adaptation. In *Human Adaptation*, ed. G.A. Harrison. Biosocial Society Series, **6**. Oxford: Oxford University Press.

Swedlund, A.C. (1978). Historical demography as population ecology. *Annual Review of Anthropology*, **7**, 137–173.

Thomas, R.B. (1976). Energy flow at high altitude. In *Man in the Andes*, ed. P.T. Baker and M.A. Little. Stroudsburg, PA: Dowden, Hutchinson & Ross.

Thomas, R.B. (1997). Wandering toward the edge of adaptability: adjustments of Andean people to change. In *Human Adaptability: Past, Present and Future*, ed. S.J. Ulijaszek and R.A. Huss-Ashmore. Parkes Foundation Workshop. Oxford: Oxford University Press.

Ulijaszek, S.J. (1997). Human adaptation and adaptability. In *Human Adaptability: Past, Present and Future*, ed. S.J. Ulijaszek and R.A. Huss-Ashmore. Oxford: Oxford University Press.

van der Eng, P. (2000). Food for growth: trends in Indonesia's food supply, 1880–1995. *Journal of Modern History*, **30**, 591–616.

Weiner, J.S. (1972). Tropical ecology and population structure. In *The Structure of Human Populations*, ed. G.A. Harrison and A.J. Boyce. Oxford: Clarendon Press.

Wilson, C. and Airey, P. (1999). How can a homeostatic perspective enhance demographic transition theory? *Population Studies*, **53**, 117–128.

Wood, J.W. (1990). Fertility in anthropological populations. *Annual Review of Anthropology*, **19**, 211–242.

Wood, J.W. (1998). A theory of preindustrial population dynamics: demography, economy and well-being in Malthusian systems. *Current Anthropology*, **39**, 99–135.

Wrigley, E.A. (1988). *Continuity, Chance and Change: the Character of the Industrial Revolution in England.* Cambridge: Cambridge University Press.

Zhao, Z. (1997). Demographic systems in historic China: some new findings from recent research. *Journal of the Australian Population Association*, **14**, 201–232.

Glossary

adaptive niche: An ecological context to which a population adapts by natural selection.

adaptive trait: A characteristic which benefits an organism's survival and reproduction.

age-specific death rate: The number of deaths in a given period to persons of a specified age divided by the population of that age in that period.

age-specific fertility rate: The number of births in a given period to women of a specified age divided by the number of women of that age in that period.

allele: Alternative form of a gene occupying the same position on a paired chromosome.

Alu **polymorphism:** DNA sequences that have been inserted at random locations in the primate genome. They are termed *Alu* because they contain a sequence recognised by the *Alu* restriction enzyme, and they are present at specific chromosome locations in some individuals but not in others.

amenorrhoea: Suppression or absence of menstruation.

antibiotic: A chemical substance that destroys or inhibits the growth of other micro-organisms, mostly used to treat infections caused by bacteria or fungi.

antimicrobial agent: Any compound that kills or inhibits micro-organisms (*see* antibiotic).

assortative mating: Choice of mating or marriage partner based on given characteristic(s).

autochthonous: Indigenous to a particular place.

autosome: Any chromosome that is not a sex chromosome.

bride price: Payment from groom or his family to bride's family to purchase the bride's labour and reproductive capacity and to repay the bride's family for the cost of raising her.

carrying capacity: Number of individuals that the resources of a habitat can support.

caste system (India): A system of socioeconomic and occupational categories established in India approximately 3,000–4,000 years ago.

cephalic index: A calculation of the ratio of width to length of skull.

classificatory: Where a term, such as 'cousin', is used for someone regarded as a cousin, although the precise biological relationship may be different.

closed population: A population that experiences no immigration or emigration.

cognitive skills: Skills which enable the individual to interpret their environment, the problems and opportunities it offers, and respond in such a way as to enhance their likelihood of survival.

cohort: A group of persons who experience some event within the same period, most commonly a group of persons born within the same period (a birth cohort).

completed family size (CFS): Cumulative fertility reached by the end of reproduction, whether of women, marriages or cohorts.

consanguinity: Genetic (generally close) relationship.

corvée: A system of labour for little or no pay in exchange for some benefit.

counter-urbanisation: The tendency for larger cities to lose population through migration to smaller places.

cross cousins: Cousins whose parents of the opposite sex are brother and sister.

crude birth rate: The number of births in a given period divided by the average population during that period.

crude death rate: The number of deaths in a given period divided by the average population during that period.

crude rates of birth and death: Numbers of births or deaths per year per 100 population members. They are misleading because they are affected by the age structure of the population (hence the proportion of people with a high probability of giving birth or dying) as well as true fertility or mortality.

***de facto*:** In actuality, the observed situation, such as the actual physical place of residence.

***de jure*:** In law, the legal situation, such as the legal place of residence.

deletion: When used in Genetics, a section of chromosome missing in some individuals.

demesne: An area of land held by a feudal lord.

demographic regimes: Combinations of interrelated demographic characteristics in a given population.

demographic transition: The process by which populations move through fertility–mortality space from the high fertility/high mortality situation characteristic of pre-industrial societies to the low fertility/low mortality situation characteristic of modern developed societies.

dermatoglyphics: The study of fingerprints.

diploid: Double or twofold; in Genetics refers to paired genes.

–genome: Complete set of paired genes.

DNA (deoxyribonucleic acid): The chief component of chromosomes, the biochemical basis of genes.

doubling time: Population doubling time in years is given by $69.3/n$, where n is the percentage annual growth rate. It is a projection, not necessarily a realistic projection.

ecumene: The permanently inhabited areas of the earth.

egalitarian: In anthropology, used to denote societies or social groups lacking ascription of social status (other than on the basis of gender).

endogamy: Marriage to a partner within a given group.

energy requirement: The amount of energy required from daily food intake to sustain all bodily functions and physical activity. Requirements vary according to the age, sex, body size and physical activity of the individual.

epidemic: An outbreak of disease, usually infectious, that spreads rapidly and widely.

epithelium: Layer of cells lining a cavity or covering the surface of an organism.

ethnic boundaries: Boundaries maintained by concepts of cultural identity.

ethnogenesis: Demographic processes leading to the establishment of current populations.

ethnographic study: Account of a society, social groups or culture based on fieldwork.

evolutionism: The use of Darwin's theory of evolution to explain other processes.

exogamy: Marriage to a partner outside of a given group.

fallow time: The time when previously cultivated fields are left uncultivated before the next period of cultivation.

famine: Extreme scarcity of food in a district or area.

feedback: The return of some part of the output of a process.

fertility: The process which produces births, or a measure of the intensity with which births are produced in a population.

fertility–mortality space: A two-dimensional space in which the coordinates indicate levels of fertility and mortality.

fertility rate: The number of live births that occur in a time period per thousand women of childbearing age.

feudal lord: A local political and administrative leader who has received authority to manage a district and its inhabitants in return for loyalty to the king.

fief: The land granted to a feudal lord by the king.

founder effect: A form of genetic drift caused by a small number of original ancestors.

gamete: Egg or sperm.

gastrointestinal tract: All parts of the digestive system including liver and pancreas.

gene: A unit of inheritance located on a chromosome.

gene flow: Migration of genes between populations.

gene genealogy: The genealogy that relates the copies of a gene present in any two or more individuals.

gene tree: Tree depicting the evolution of a particular gene.

genetic distance: A measure of the extent of difference in genetic variation between two (or more) populations.

genetic drift: Changes in the gene frequencies of a population due to chance factors.

genetic structure: The pattern of genetic variation in a population.

genome: A complete set of genes; can be used for a species, a population or an individual.

genotype: The genetic constitution and condition of an organism.

globalisation: The process enabling markets to operate internationally.

guerrilla: A fighter in a campaign of war who avoids pitched battles with the enemy and instead undertakes small, unexpected assaults on the enemy or his property.

haploid (genetics): Half the number of genes, one of every pair.

heterozygote: An organism in which each of the pair of alleles referred to are different.

HLA (Human Leucocitary Antigens): Set of proteins found in the membrane of white cells, extremely variable among individuals, and used by the immune system to distinguish 'self' from 'non-self'.

homozygote: An organism in which each of the pair of alleles referred to are the same.

household: The people who share a single cooking fire or kitchen and live under one roof.

Human Genome Project: International scientific project aiming to decipher the human genome.

hypergamy: Marriage with one belonging to a higher socioeconomic group or caste.

hypermobility: Intensive or excessive mobility and migration.

immune system: System responsible for resistance against infection, comprising antibodies and white blood cells.

indigenous: A community native to the region in which they live; in contrast to colonist or settler populations who arrived later in the same region.

infant mortality rate (IMR): The number of deaths of infants under one year of age per 1,000 live births in a given year (*see also* neonatal deaths).

infanticide: The deliberate killing of infants.

isogamy: Marriage between two poeple belonging to the same socioeconomic group or caste.

isogrowth lines: Lines joining points with equal rates of population growth in fertility–mortality space.

least developed countries (LLDCs): The poorest countries, mostly found in sub-Saharan Africa.

less developed countries (LDCs): The poorer countries of Africa, Asia and Latin America, sometimes called developing.

levirate marriage: Marriage of a younger brother to his deceased older brother's widow.

life expectancies: Average years lived by a defined population or set of individuals under defined circumstances, calculated from birth unless otherwise specified. Low life expectancy usually implies large numbers of deaths in infancy and early childhood.

lineage: A line of biological or social inheritance.

macro-state: A large state, sometimes defined as having more than 100 million inhabitants.

matrilocal marriage: Marriage after which the couple live in the home of the bride's mother.

mega-city: Very large city, now often defined as having more than 10 million inhabitants.

menarche: The onset of menstruation at puberty.

Mendelian population: A population defined by reproductive patterns, in which any individual is most likely to mate with another individual from the same population, thereby forming a common gene pool.

micro-state: A small state. Sometimes defined as having fewer than 1 million inhabitants.

migration: Movement from one place to another involving a permanent or semi-permanent change of residence.

millionaire cities: Cities with more than a million inhabitants.

mitochondrial DNA (mtDNA): Small piece of DNA found outside the chromosomes and within the mitochondria. Since sperm mitochondria do not enter the egg in fertilisation, mtDNA is transmitted solely through the female line.

mitochondrion (pl. mitochondria): Each part of a cell that is enclosed by a double membrane and that produces most of the energy needed by the cell.

mobility: Movement.

–social: Movement through socially defined groups.

–spatial: Movement over geographic space.

modern synthesis (of evolution): Theory encompassing both the findings of molecular biology and the theoretical developments about evolution.

molecular: Relating to molecules.

momentum: In demography, the increase (or decrease) in population size which would occur if the fertility of a population changed immediately to the level which would just ensure the replacement of each generation.

morbidity: Condition of ill health or disease.

morbidity rate: Number of cases of a disease found to occur in a specified number of the population, usually given as cases per 100,000 or per million.

more developed countries (MDCs): Richer countries of Europe, North America, Japan and Australasia.

morphology: The study of structure and form of an organism; or the form itself.

morphometrics: Measurements of morphological features such as head length, arm length, nose width, etc.

mortality: The process which produces deaths, or a measure of the intensity with which deaths occur in a population.

mortality rate: The incidence of death in the population in a given period, e.g. annual mortality rate is the number of registered deaths in a year, multiplied by 1,000 and divided by the mid-year population size.

multiple birth: Birth of twins, triplets or higher order multiples.

mutation: Any alteration of the DNA code.

–somatic: Occurring in any non-germ-line body cell.

–germ-line: Occurring in any of the cells that, after a number of divisions, will produce a gamete.

natural fertility: Fertility realised in the absence of deliberate birth control: the prime criterion of control is that a couple modify their reproductive behaviour to a target family size.

natural increase: The difference between the number of births and the number of deaths in a population (if deaths exceed births, it is referred to as natural decrease).

neonatal death: Death of an infant within 28 days of birth.

net migration: The difference between the number of immigrants and the number of emigrants in a population.

net reproduction rate: The ratio between the size of the next generation and the size of the current generation. It is approximately equal to the total fertility rate multiplied by the proportion of births which are girls multiplied by the probability of a woman surviving to the mean age of childbearing.

neutral markers: Identifiable section of DNA that is neither selectively positive nor negative.

non-coding regions of DNA: Regions of DNA that do not code for any known protein.

non-recombining portion of Y chromosome: The portion of the Y chromosome that does not exchange material with the X chromosome during meiosis.

nuclear DNA: DNA in the nucleus of the cell.

nuclear family: A family made up of parents and their unmarried children.

nucleotide: Each of the four basic chemicals (usually abbreviated A, C, G, T) that constitute DNA.

nucleotide sequence of DNA: A sequence of the component particles of DNA.

nuptiality: That which pertains to marriage.

nutrient requirement: Quantity of various components of the diet (e.g. energy, protein, iron, calcium) needed to maintain health (*see* recommended nutrient intake).

oral rehydration therapy: A solution of sodium and glucose (or salt and sugar) in water administered to patients with diarrhoea, thus preventing dehydration.

organism: Any living, or once-living, being.

over-population: A higher level of population than can be sustained by a region's resources under current methods of production.

Painted Grey Ware: A type of pottery.

palaeoanthropology: The study of early humanity and their ancestors.

parish records: Registers kept by a church administrator of baptism, marriage, and burial services conducted at that church.

partible inheritance: Inheritance in which the property of a household is divided between all heirs (usually the children).

participant observation: A method of data collection in which the observer participates more or less completely in the activities of the people who are being observed.

pathogen: Any agent of disease, usually microscopic.

patrilineal: Through the lineage of male ancestors and descendants.

patrilocal marriage: Marriage after which the couple live in the home region of the groom's father.

patronage: Resources provided by a patron.

peasant: A cultivator whose productive activities are primarily directed towards his or her household's subsistence needs, but who tends to have obligations in a wider society.

perinatal death: Death taking place in the perinatal period, from 24 weeks after conception until 7 days after birth, including stillbirths and live-born infants.

period: A given span of calendar time, for example the period 1995–1999.

peri-urban development: Extensions of the built environment around the fringes of towns and cities.

phenotype: Characteristics or condition of an organism due to genetic and non-genetic influences.

polyandry: Marital custom where more than one husband is allowed.

polygyny: Marital custom where more than one wife is allowed.

polymorphism: Where more than one form of an allele exists in a population with a frequency that exceeds 1%.

population: A group of persons who can be delimited on the basis of some observable characteristic.

population mobility: All phenomena involving the displacement of individuals.

population redistribution: Changes in the distribution of population through migration or natural change.

population structure: The distribution of various characteristics across the members of a population.

pre-Darwinian: Before the writings of Charles Darwin.

premature birth: Birth before full term; more specifically, pre-term birth refers to all births before 37 weeks gestation.

primary sector: Economic activities involving agriculture, fishing, forestry, hunting and mining.

primogeniture: Inheritance in which the property of a household passes intact to the oldest child.

protein: A compound formed by a chain of amino acids.

rate: In Demography this refers to the number of persons experiencing an event in a given period divided by the number of persons exposed to the risk of experiencing that event during that period (e.g. *age-specific fertility rate, age-specific death rate, crude death rate, crude birth rate*).

rate of natural increase: Crude birth rate minus crude death rate, i.e. annual rate of population growth if migration is disregarded.

reciprocity: Equal exchange between two partners, in which each partner takes it in turns to give something to, or perform a service for, the other.

recommended nutrient intake: Level of intake of a specified nutrient that is required to maintain health in individuals. The level required is dependent on age, sex, body weight, physical activity and the nature of the diet.

replacement level: The level of fertility required in a population so that each generation is just replacing itself (i.e. the net reproduction rate is 1.0, and the population growth is zero).

rural depopulation: Loss of people from rural areas, usually to urban areas or abroad.

secondary sector: Economic activities involving the manufacture of goods from raw materials.

seignioral: Rights and duties of a feudal lord.

selection: Used in biology for the idea of natural selection.

–balancing: Where two (or more) alleles co-exist and continue to co-exist due to a balance of selective forces, e.g. that favouring the heterozygotes over both homozygotes.

–positive (advantageous): Where the allele or characteristic is selected for and increases in frequency.

–negative (purifying): Where the allele or characteristic is selected against and decreases in frequency.

sex chromosomes: The chromosomes, X and Y, which determine the sex of an organism. [In humans and all other mammals, females carry two X chromosomes and males carry one X and one Y chromosome.]

sex-specific migration: Migration of only one of the two sexes.

single nucleotide polymorphism (SNP): Allelic forms that differ by exactly one nucleotide.

social change: Change in the structure of social relationships.

stable population: A closed population in which fertility and mortality have been constant for a long time, leading to a constant crude birth rate and a constant set of age-specific death rates.

stationary population: A stable population in which the total number of births equals the total number of deaths, so that the growth rate is zero.

stillbirth: Birth of a foetus, later than 24 weeks after conception, that shows no signs of life at any time following the birth.

subsistence farming: Productive activities primarily directed towards needs of own household without participation in the market.

sudden infant death syndrome (SIDS): Also known as cot death; the sudden death of a baby, often overnight, from an unidentifiable cause.

swidden: A practice based on cutting down and burning forest to create temporary space for cultivation, also known as 'slash-and-burn' farming.

synergism: The interaction of two agents to provide an effect which is greater than the sum of the effects when the agents act separately.

tandem repeat polymorphism: A sequence of DNA that is repeated multiple times in tandem (e.g. CAGACAGACAGA).

tertiary sector: Economic activities involving services of all kinds.

total fertility rate: A measure of the average completed family size of a cohort of women. It is the sum of the age-specific fertility rates over the childbearing age range (normally 15–49 years).

unigeniture: Inheritance in which the property of a household passes intact to a single child (often to the oldest, i.e. primogeniture).

urbanisation: Increase in the proportion of population living in urban areas.

uxorilocal: In the natal locality of the wife.

vigilante patrols: Volunteer law enforcement patrols organised by members of a community.

villein: A person who is the tenant of a feudal lord; in return for land to cultivate, the villein must perform services for the lord and accept his authority.

virilocal: In the natal locality of the husband.

wasting: A form of malnutrition defined by a measure of weight adjusted for age (weight-for-age value) less than 2 standard deviations below the reference mean weight at that particular age. The international reference values commonly applied are those devised by the World Health Organisation and Centres for Disease Control.

X chromosome: Sex chromosome, when paired responsible for the female sex of a mammal.

Y chromosome: The sex chromosome responsible for the male sex of a mammal. Only males carry and transmit it.

Index

CPSIA information can be obtained at www.ICGtesting.com
Printed in the USA
BVOW011648171111

276327BV00001B/50/P